Your Cannabis CBD:THC Ratio

Your Cannabis CBD:THC Ratio: A Guide to Precision Dosing for Health and Wellness
By Uwe Blesching, Ph.D.

Published by Quick American
A Division of Quick Trading Co.
Piedmont, CA, USA

ISBN: 9781936807482
eISBN: 9781936807536

Printed in the United States
First Printing

Special thanks to the team that made this book a reality:

Agent: Doug Reil
Art Director: Christian Petke
Project Manager: Christy Quinto
Publisher: Ed Rosenthal
Project Director: Jane Klein
Graphic Design: Scott Idleman / Blink
Editor: Kathy Glass
Editor: Matthew Hoover
Acquisition Editor: Rolph Blythe
Cover Design: Tabitha Lahr
Proofreader: Paula Dragosh
Graphic Vitruvian Wo/Man: Tim Sunderman

Library of Congress Control Number: 2020941050

Your Cannabis CBD:THC Ratio

A Guide to Precision Dosing for Health and Wellness

by Uwe Blesching, Ph.D.

Praise for *Your Cannabis CBD:THC Ratio*
A Guide to Precision Dosing for Health and Wellness

"Uwe Blesching is one of our new wisdom carriers, bringing modern science to Mother Nature's ancient gift, teaching us how to get the very best from our cannabis consumption. When I have a question about the biochemistry of cannabis and don't know the answer, I call Uwe first."
  —**Steve DeAngelo,** Co-Founder of Harborside and author of *The Cannabis Manifesto: A New Paradigm for Wellness*

"An excellent current review of cannabis therapeutics from an expert perspective."
  —**Michael Backes,** author of *Cannabis Pharmacy: The Practical Guide to Medical Marijuana*

"An indispensable book for any clinician working with cannabis patients. It has already led to better cannabis recommendations and better outcomes at my clinic."
  —**Martha Montemayor,** Director, Cannabis Clinicians Colorado; Producer, Marijuana for Medical Professionals Conference

"A clear and well-researched guide to cannabis as a medicine that will be vital for both patients and their caregivers. Blesching has done a masterful job of including and supporting some of the most vital information that patient and practitioner want, but few have been able to obtain."
  —**Maria Mangini,** Ph.D., FNP, CNM

"What Uwe has done is show the promise of cannabinoid treatments for various diseases based upon a combination of pre-clinical, clinical (where available), and anecdotal evidence that offers guidelines as a starting point when seeking self-treatment, and when opening dialogue with a medical provider. I applaud the emphasis on the importance of THC and other cannabinoids, dispelling the myth of CBD being the 'medical' part of the plant. An excellent reference for your bookshelf!"
  —**Mara Gordon,** Co-founder of Aunt Zelda's and Zelira Therapeutics

"Due to the diversity of bioactive cannabinoids and terpenoids present, the biomedical science of cannabis is extraordinarily complex. This book provides an excellent map of the existing terrain."
  —**David Presti,** Ph.D., Teaching Professor of Neurobiology, University of California, Berkeley, and author of *Foundational Concepts in Neuroscience: A Brain-Mind Odyssey*

"Over the last 30 years, millions of people have used cannabis to alleviate symptoms and to heal. However, patients have had to guess and experiment to find the most efficacious ratios of the different cannabinoids. Uwe Blesching provides information on how to use cannabis in ways that help the most, empowering patients to heal themselves with this safe and gentle herb."

—**Ed Rosenthal,** Cannabis author and researcher

"Gives people the understanding of cannabinoid ratios to use on their journey to health— this is a gift that all deserve."

—**Kimberly Cargile,** CEO, A Therapeutic Alternative

"*Your CBD:THC Ratio* is like a sequel to the author's amazing book, *The Cannabis Health Index.* It gives us an advanced understanding of the plant's many compounds and its usages for individuals. Uwe Blesching remains a Cannabis Sadhu not just for our generation but for coming ones too."

—**Priya Mishra,** Hempvati (India's first female cannabis activist)

"Uwe Blesching is a creative genius of cannabinoid health science education and communication. No one is able to distill the complex and varied science of cannabis for integrative human health in a practical and usable way like he has. This book is a powerful reference and milestone in deepening our understanding of the human–cannabis relationship and for empowering readers in their ability to heal with cannabis."

—**Sunil Aggarwal,** M.D., Ph.D., FAAPMR, co-founder and co-director of the Advanced Integrative Medical Science Institute and Affiliate Assistant Professor of Medicine and Geography at the University of Washington

"Uwe Blesching has created a masterpiece of understanding and education on cannabis, the science of THC and CBD, and how the plant works with our bodies to keep us well. Uwe writes in such a way that the average person can easily understand the seemingly complicated science of this superfood that is cannabis."

—**Sharon Letts,** Writer at *High Times,* Producer

"Once more, Blesching has hit the nail on the head. The guide that everyone has been waiting for."

—**Michael Green,** Cannabis Libris

About the Author

Uwe Blesching is a medical journalist and regular contributor to the cannabinoid health sciences, mind-body medicine, phytopharmacology, as well as evidence-based illness prevention and treatment protocols. In addition to his lifelong passion for integrative medicine, he holds an MA in Psychology and a PhD in Higher Education and Social Change from the Western Institute for Social Research. His writings are informed by his rigorous in-depth research and twenty years of experience in emergency medicine as a paramedic for the City of San Francisco.

His other books are *The Cannabis Health Index* and *Breaking the Cycle of Opioid Addiction: Supplement Your Pain Management with Cannabis*. Uwe Blesching lives in Berkeley, California. Visit his website at www.uweblesching.com.

Contents

Introduction

Why Cannabis Works

People use cannabis for all sorts of reasons. Whether you are here to treat a passing symptom, a chronic condition, a stubborn manifestation of what ails you—or even if you are here to develop more agency about what kind of cannabis experience you wish to have—this book can help you heal in two ways. One, it provides you with evidence-based information to make more discerning decisions and informed choices about what type of cannabis to use; and two, it provides you with a step-by-step process of concrete actions you can take to achieve and sustain the precise effects you want, need, or desire.

After a decade of doing research in this field, I have come to believe that cannabis works for so many different types of patient populations because it produces changes in the body as well as the mind in a way that supports and fortifies the body's capacity for self-healing. In a nutshell, here is why. There are eleven organ systems in the human body, and cannabinoid receptors are present in all of them. When cannabis binds with these naturally occurring receptors, people benefit because cannabis is the ultimate *adaptogen*, a natural substance that gives each system what is needed to thrive.

From historical use and modern research, we know that cannabis consumption can benefit patients suffering from many different medical problems. A quick search for the words *cannabis* and *marijuana* on PubMed, the largest medical repository of scientific studies in the world, yields tens of thousands of relevant studies. And a good number of them offer direct insights specific to the needs of chronic patients dealing with stubborn symptoms. For instance, the sidebar on page 10 includes a list of chronic conditions for which there is no cure within the orthodox treatment paradigm. The best that most of these patients can hope for is managing their symptoms with expensive pharmaceutical drugs. And, as many of these patients very well know, long-term use, even if it provides some temporary relief, often comes with an additional price tag in the form of a number of adverse effects. It ought to come as no surprise that more and more people are turning to more natural, effective, safer, and less expensive alternatives. This chronic patient-driven trend is not just a fluke produced by the lack of progress within Western medicine but is also supported by the evidence. Once again, consider the conditions listed in the sidebar. While a more complete list to date contains over 250 conditions for which cannabinoids may show therapeutic potential, it includes only those conditions for which there are at least five scientific papers/studies that indicate degrees of evidence (possible, probable, and actual) that using cannabis may be beneficial.

Helpful Terms to Understand for This Chapter

Cannabis produces a great number of biologically active compounds, and THC and CBD tend to occur most abundantly. Both are known to be able to induce a vast variety of therapeutic effects relevant to a great number of different patient populations.

THC: Tetrahydrocannabinol is the primary psychoactive cannabis constituent.

CBD: While cannabidiol is generally considered a non-psychoactive cannabis constituent, it is fully capable of producing measurable affectual changes or mood improvements.

Chemotype: In this book a chemotype of cannabis is primarily defined as the ratio between THC and CBD.

In the years of doing research for my books and decades of working in the Emergency Medical Services (EMS), I have come to know the value of two critical approaches that can help us heal and guide the structure of this book. First, the currently available evidence is not perfect (because it is ever evolving), but it is essential for guiding the best possible treatment we have to date. And second, having a practical, patient-centered, concrete way of realizing the scientific evidence is the smart way to go.

This is certainly true in EMS, where treatment algorithms (step-by-step pathways) adjust every two years or so as new evidence becomes available, sometimes upending practices previously taken for granted. For instance, as every health care provider and person certified in cardiopulmonary resuscitation (CPR) knows, by the time recertification comes around, you will likely learn a new set of compression ratios and number of respirations to perform in this basic life-saving measure. And new ones again in the future. However, the fact is, in nearly all other fields of modern medicine, standards of care are not based on the latest evidence. There is no guarantee that you will receive the best possible care based on the latest evidence.

The good news is that when it comes to cannabis medicine, we do not have to wait for the medical-industrial complex to catch up and translate the evidence into practical treatment protocols. We have the ability to do it ourselves. This freedom of use, however, brings with it the responsibility to learn as much as we can about cannabis as medicine, in order to maximize therapeutic results, and of course minimize adverse effects.

A Typical Cannabis Patient's Experience

When seeking relief, most people understand that Vicodin, Xanax, Zoloft, and Ambien are all medicines, but they do not give you the same results. Each product has a uniquely different chemical makeup and thus will produce very different effects: Vicodin is for pain; Zoloft for depression; Xanax for anxiety; and Ambien for sleep. Virtually every pill prescribed by your physician has a unique name, purpose, standardized chemical composition, list of possible adverse effects, and specific instructions for use.

As you will learn in the following pages, the same is true for the three basic chemical variations of cannabis—one type may be optimal for chronic pain conditions (such as central pain from brain

Scientific Studies Show Therapeutic Potential of Cannabis for Many Chronic Conditions

Acute pain (20 studies)

Alcohol Dependence, Withdrawal, and Intoxication (13 studies)

Alcoholic Liver Disease (5 studies)

Allergic Contact Dermatitis (6 studies)

Alzheimer's Disease (22 studies)

Amyotrophic Lateral Sclerosis (12 studies)

Anorexia Nervosa (16 studies)

Anxiety and Panic Disorders (59 studies)

Arthritis (11 studies)

Asthma (22 studies)

Atherosclerosis (7 studies)

Attention Deficit and Hyperkinetic Disorders (8 studies)

Autism and Pervasive Developmental Disorders (15 studies)

Bacterial Infections (18 studies)

Bipolar Affective Disorder (8 studies)

Bone Cancer (6 studies)

Brain Cancer (30 studies)

Breast Cancer (37 studies)

Cachexia (6 studies)

Cancer of the Pancreas (8 studies)

Cancer Pain (14 studies)

Cardiac Arrhythmias (15 studies)

Central Pain Syndrome (5 studies)

Cervical Cancer (6 studies)

Chronic Obstructive Pulmonary Disease (9 studies)

Chronic Pain (48 studies)

Cocaine Dependence (7 studies)

Cognitive Dysfunction (23 studies)

Colon Cancer (19 studies)

Cystitis (5 studies)

Depression (30 studies)

Diabetes Mellitus (27 studies)

Dravet Syndrome (17 studies)

Dysbiosis/Microbiome Imbalance (6 studies)

Dystonia & Related Conditions (5 studies)

Encephalitis, Myelitis, and Encephalomyelitis (7 studies)

Epilepsy (64 studies)

Fatty Liver Disease (10 studies)

Fibromyalgia (7 studies)

Fragile X Syndrome (9 studies)

Glaucoma (16 studies)

Heart Attack (8 studies)

Heart Disease (31 studies)

Hepatitis (10 studies)

Herpes (10 studies)

HIV/AIDS (22 studies)

Huntington's Disease (23 studies)

Hypertension (17 studies)

Intermittent Explosive Disorder (5 studies)

Lennox-Gastaut Syndrome (19 studies)

Leukemia (12 studies)

Lymphoma (9 studies)

Melanoma (11 studies)

Metabolic Syndrome (8 studies)

Metastatic Cancer (16 studies)

Migraine (13 studies)

Multiple Sclerosis (63 studies)

Nausea and Vomiting (43 studies)

Neurodegenerative Disorders (38 studies)

Nicotine Dependence and Withdrawal (16 studies)

Obesity (9 studies)

Obsessive-Compulsive Disorder (5 studies)

Opioid Dependency and Overdose (49 studies)

Organ Transplant/Graft Rejection/Failure (6 studies)

Osteoporosis (9 studies)

Palliative Care (20 studies)

Pancreatitis (14 studies)

Parkinson's Disease (28 studies)

Peripheral Neuropathy (23 studies)

Post-Traumatic Stress Disorder (25 studies)

Prostate Cancer (12 studies)

Psoriasis (5 studies)

Psychosis (41 studies)

Reperfusion Injury (10 studies)

Rheumatoid Arthritis (22 studies)

Schizophrenia (31 studies)

Sepsis (10 studies)

Severe Stress Reaction/Oxidative Stress (33 studies)

Sickle Cell Disorders (8 studies)

Skin Cancer (Non-Melanoma) (8 studies)

Spinal Cord Injuries (24 studies)

Stomach Cancer (6 studies)

Stomach Ulcers (12 studies)

Stress and Life Management Difficulty (58 studies)

Stroke/Cerebral Infarction (34 studies)

Thyroid Cancer (5 studies)

Tic Disorders and Tourette's Syndrome (16 studies)

Traumatic Brain Injury (17 studies)

Vascular Dementia (5 studies)

or spinal cord injuries), though it also produces the most significant changes in cognition, in other words, it will get you "high." In contrast, another basic type of cannabis is optimal for treating seizure-related disorders while producing no changes in cognition—it will not get you "high." Yet a third type of cannabis engenders effects that lie somewhere in between, significant for many different patient populations such as those with multiple sclerosis.

Yet when your physician gives you a cannabis recommendation (enabling you to legally purchase cannabis in states that permit medical marijuana), you generally receive a prescription for just cannabis—not any particular variety, of which there are hundreds. You are left on your own to figure out "which pill to buy." Patients and consumers have no evidence-based way of knowing which of the many options will work for them and which to avoid.

Upon entering a well-stocked dispensary, today's patient is presented with hundreds of jars of cannabis flowers from a plethora of growers; edible cannabis in the form of chocolate, candy, or cookies; tinctures based in glycerin, alcohol, or oils. There are concentrates (typically CO_2 or solvent-based); there are various types of hash, live-resin, and kief; topical preparations in various forms (creams, salves, time-release patches); capsules; oral or sublingual sprays; vaginal or rectal suppositories, and more.

How is anyone to determine which product and consumption method is right for them? How will they take into account the big differences in onset time and effects produced by inhaling versus ingesting? Patients who regularly visit a dispensary eventually learn to navigate this complex setting, but even the savviest cannabis users acknowledge how confusing the variety of products can be. With Tylenol and Sudafed, you can at least read and compare their labels, which contain consistently presented information.

Consider the experience of Maureen Dowd, a columnist for *The New York Times*.[1] Dowd thought it would be interesting to write from personal experience about the "green revolution" spreading across the nation, indeed the world. On a visit to Denver, Colorado, where even recreational cannabis is now legal, she purchased a caramel-chocolate-flavored candy bar. She took a nibble in her hotel room, but nothing happened. She nibbled some more… After an hour or so she "felt a scary shudder go through my body and brain".

Her unpleasant experience seemed to progress to a full-blown panic attack that left her curled up on her bed paranoid for "the next eight hours," unable to reason or function. It would have been helpful for her to know that unpleasant adverse effects can be easily abated or avoided—by creating a relaxing environment (e.g., playing gentle music), and through the presence of a trusted friend who provides calming reassurance. And perhaps she did not know that different varieties of cannabis, and even methods of using it, produce profoundly different effects on people.

Dowd is not alone in her experience. Sadly, it happens every day. It is especially tragic when patients, suffering from stubborn symptoms and chronic conditions, reject the use of cannabis because of common beginner's mistakes, and the lore surrounding them. These patients forgo cannabis's therapeutic potential.

Cannabidiol, the New Superfood

CBD is one of the compounds that tend to occur in cannabis abundantly along with THC. While people in countries and states with legal access to cannabis tend to be more familiar with the therapeutic and wellness potential of this plant-based compound, many people are not. If Maureen Dowd had used a cannabis-based product abundant in CBD, her experience would have been very different. First of all, she would not have felt "high" or anxious; instead, she would have likely felt a general uplift in her mood while realizing gentle shifts and therapeutic benefits in nearly all organ systems, such as anti-inflammatory, anti-oxidative, and neuro-protective effects.

Indeed, CBD has quickly become a major player in the areas of health and healing, as well as an important addition to the nutritional and wellness market. In fact, CBD is now a proven and potent healing agent with relevance to a great number of different patient populations—such as in cases of pediatric epilepsy that do not respond to pharmaceutical treatments. CBD is acknowledged by the regulatory and economic powers that be and now exists as a pharmaceutical version available by prescription, fully recognized and approved by the Federal Drug Administration. However, the big-pharma price tag is high. An annual supply can cost over thirty thousand US dollars. Compare that to the fact that you can grow a tall CBD plant in your garden for nearly nothing.

CBD is not limited in its potential to treat a number of chronic conditions but has also found its way into tinctures and oils used effectively in people and animals alike. Oils have been shown to produce significant beneficial effects such as reducing certain types of pain such as chronic pain, and CBD has been noted has been found to significantly reduce the pain and suffering of dogs experiencing arthritis. A great number of nutritional and wellness products promote improving in memory, reducing free radicals, or mitigating the effects of aging (to name but a few) with CBD as the primary active ingredient. (More on CBD in chapter 3.)

It Doesn't Have to Be This Way

Patients need to know how to achieve desired therapeutic effects and how to sustain them. The current industry-wide system for choosing cannabis or cannabis-based products is part of the problem. Our system is based on the assumed therapeutic qualities of cannabis's prime species, sativa and indica. And it is woefully imprecise.

You may ask, why imprecise? The broad qualities of these prime species of cannabis have served to inform choices for decades. Consider this: Two people use identical seeds to grow a specific cannabis species—but even subtle changes in their growing environments can significantly change the chemical composition of all plant constituents. Environmental signals such as temperature, nutrition, or quality of light will initiate specific changes in genetic expressions without changing the plant's genome or DNA. Thus, even though the two cannabis species are genetically identical, the therapeutic (or adverse) effects they engender could be as different as mixing up a Zoloft with a Vicodin.

The good thing is that we are developing better systems. As the last four decades of research have discovered, three basic variables in chemical composition are the key to realizing specific therapeutic effects from any cannabis plant: the first variable is the *amount* of the primary psychoactive cannabis constituent tetrahydrocannabinol (THC for short); second is the *amount* of the non-psychoactive cannabis constituent cannabidiol (CBD); and the third variable is the *ratio* of THC to CBD. These three variables determine the three basic types of cannabis, which are referred to as cannabis chemotypes, and the Roman numerals I, II, and III are used to identify them. We will be using honeycomb images throughout the book to give you an easy visual reference to let you know at a glance which cannabis type we are talking about.

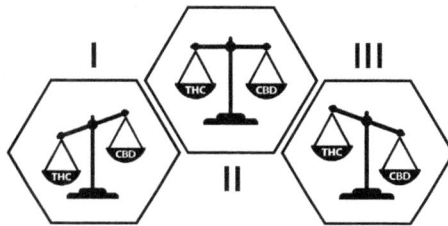

Introduction, Graphic 1: Cannabis Chemotypes

Understanding Cannabis by the Numbers: Chemotypes

Let me give you two practical examples of what you can do with understanding cannabis by these numbers: One of the most frequent concerns I hear about using cannabis is related to fears about getting "high." While I think there is nothing wrong with getting high—and for many people getting high is part of the healing process—did you know there is a type of cannabis that will induce a great number of therapeutic effects such as giving you a gentle uplift in your emotions but without any changes in cognition (without getting you high)? By understanding the three basic cannabis chemotypes, you get to determine what kind of cannabis experience you would like to have. The second most common question I am asked is, what strain is good for what condition? Being familiar with the three basic cannabis chemotypes allows you to lean on an abundance of scientific studies that have determined what effects are likely to be produced by each cannabis chemotype.

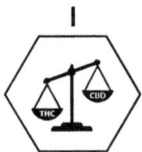

Chemotype I
Contains more THC than CBD (THC>CBD)

Take a look at some of the effects that have a chemotype I such as Godfather Og (~34% THC) or Super Glue (~32% THC) will produce: deep relaxation, conscious sedation, changes in cognition ("feeling high"), analgesia (pain relieving), mood improvements, increased appetite, and anti-tussive (cough relieving). However, this chemotype is also associated with a negative "high" experience related to anxiety or panic attacks. A chemotype I has the potential for other adverse effects such as making the very condition you would like to treat worse (e.g., insomnia, pain, or nausea and vomiting). And it is this type of cannabis that is associated with the potential for psychological addiction. Conditions for which a chemotype I is commonly indicated include HIV/AIDS-based painful neuropathies, central pain (e.g., stroke), chemotherapy-induced nausea and vomiting, severe physical withdrawal symptoms (e.g., heroin dependence), or Alzheimer's disease with night-time agitation.

Chemotype II
Contains relatively equal amounts of THC and CBD (THC=CBD)

Take a look at some of the effects a chemotype II such as LT Fire (THC ~11%:CBD ~11%) or Cellar CBD (~7.3% THC:~7.7% CBD) will produce: relaxation (but without or with very limited sedation), milder changes in cognition ("feeling high"), mood improvements, and antispasmodic effects (reduced muscle spasms). This chemotype reduces all the potential risk of a chemotype I but still engages the therapeutic potential of THC. In fact, a chemotype II can produce a synergy of therapeutic effects associated with both THC and CBD while mitigating the adverse effect potential of THC. Conditions for which a chemotype II is commonly indicated include multiple sclerosis (calms and reduces primary and secondary symptoms), cancer-based or cancer-induced peripheral neuropathies, or fibromyalgia (systemic analgesia).

Chemotype III
Contains less THC than CBD (THC<CBD)

Take a look at some of the effects a chemotype III such as Cannatonic (0.8% THC:20.8% CBD) or AC/DC (0.8% THC:15.4% CBD) will produce: mood improvement (relevant to reducing anxiety, depression, psychosis), analgesic effects (especially in synergy with opioids), anti-seizure effects (especially in cases of pediatric conditions not responsive to pharmaceutical regimens). This chemotype of cannabis produces changes in affect (mood), not in cognition, and will not get you "high" (unless you are using a strain with a relatively high amount of THC), and a chemotype III (with no or little amounts of THC) has no addiction potential. Conditions for which a chemotype III is commonly indicated include (but are not limited to) acne, heart disease, colitis, anxiety (e.g., PTSD, OCD, social phobias), depression, or a number of pediatric seizure disorders that do not respond to pharmaceutical anti-epileptics.

This Book and the CannaKeys Process

The material in this book is designed to help you help yourself, or together with your health care provider, to unlock the ultimate combination of cannabis-based constituents that are optimal for you. We term this step-by-step method the CannaKeys Process, which is as basic as filling out a medical history form at the doctor's office.

By the time you are done, you will have more than just a generic prescription for cannabis—you will know which type of cannabis is best for you.

By understanding how differences between these three basic numbers (amount of THC, amount of CBD, and the ratio between them) play out in practical terms in human biology, you will have evidence-based information allowing you to control cannabis effects in yourself or for those in your care. You will be able to precisely produce the symptom reduction and mood improvement that you are looking for. You will be able to avoid adverse effects. You will be able to calculate the optimal dose that is right for you or your patients. You will learn how to navigate the different effects that different forms of cannabis or cannabis-containing products engender and how to use them safely and effectively. Also, you will be able to fine-tune the process by creating a synergy (entourage effect) with other plant constituents such as terpenes. And, finally, you will learn how to look inward to monitor and record your progress to further fine-tune your movements on the pathway to optimal health and well-being.

This book takes you step-by-step through the process and informs you about the variables that matter (and why). First we will look at the top 10 common medical conditions for which patients go to the doctor's office and at the evidence that suggests how each might benefit from healing with cannabis. Then we will dive into the plant itself, demystify the confusion surrounding it, and explain how it affects our bodies (sneak preview: humans are hardwired to respond to cannabis). We will illuminate the differences among varieties of cannabis, teach you how to know what type is best for you, and then work through the CannaKeys Process.

Thoughout this book you will see the repeated use of honeycomb-like graphics with a scale inside (THC on the left side of the scale and CBD on the right) that will tell you at a glance which type of cannabis we are talking about.

Key Points, Introduction: Why Cannabis Works

- THC and CBD tend to be the most abundant compounds produced by the plant cannabis.
- Considering the far-reaching therapeutic potential and the possible risks of cannabis produces a favorable risk versus benefit conclusion.
- Working with cannabis chemotypes can provide you with agency about what kind of cannabis experience you wish to have.
- Using cannabis chemotypes allows you to make more informed and empowered choices that are grounded in the currently available scientific evidence.
- The book and the CannaKeys Process will help you make empowered choices to produce and sustain the precise effects you desire.

Chapter 1

Top 10 Reasons People Seek Medical Care, and the Appropriate Type of Cannabis for Each

In this chapter you will learn about cannabis treatment for the following:
- Cancer
- Heart Disease
- Chronic Pain
- Diabetes
- Anxiety
- Neurological Conditions
- Insomnia
- Colitis
- Cough
- Skin Conditions

In the introduction we saw a long list of conditions for which cannabis may hold promise as a therapeutic agent. In this chapter we will look briefly at some specific, common conditions—those for which people need the most help—and how cannabis is thought to work. Each of the 10 conditions is presented as a summary based on a review of select available literature. Readers who would like more details can consult the Appendix: The Science Behind the Top Ten.

The Mayo Clinic, a medical research and practice group considered one of the best in the United States,[1] conducted a study about the most common reasons people go to the doctor. Over 140,000 patients participated,[2] and the resulting top 10 were skin disorders (e.g., acne); chronic (non-malignant) pain from the back, joints, and head; upper respiratory diseases (e.g., cough); mental disorders (e.g., anxiety); neurological illness (e.g., multiple sclerosis); heart disease; diabetes; cancer; gastro-intestinal conditions (e.g., IBD); hypertension; and disorders of lipid metabolism (atherosclerosis).[3]

While the point was already made in the introduction, it is perhaps worth repeating that within the orthodox treatment paradigm, many of these conditions are considered chronic, that is, without hope for a lasting cure. The primary treatment is often reduced to the management of symptoms with pharmaceuticals, which often come with a considerable price tag and the very real risk of serious adverse effects. Our need for cheaper, safer, and deeper healing remains. It is no surprise that more and more patients are exploring the cannabis alternative.

Whether you are using cannabis for chronic illness or a more acute condition, the information that follows is designed to give you an overview of how cannabis is currently being employed (with success) to treat a range of specific conditions. We will explain the framework of each condition and give you a summary on how cannabis is thought to mitigate ill effects and provide relief.

To give you the benefit of practical examples up front, we are asking you to jump right in—not all the terms used may be familiar to you, but we define most of the relevant ones in sidebars. Terms will be explained later in greater detail, so you do not have to completely grasp all the new concepts immediately.

The reader is reminded that while the following evidence-based suggestions may serve as a good starting point to determine which type of cannabis may be best suited for these conditions, it is only one variable (albeit a significant one) that determines what kind of effect each type will have on your body, mind, and emotions. As you read through this book, you will become acquainted with the step-by-step CannaKeys Process that shows you in much greater detail how to address all of cannabis's controllable variables, so you can realize and sustain the precise effects you are looking for.

Helpful Terms to Understand for This Chapter

Cannabinoid: A chemical compound that interacts with the human body, producing a measurable effect in body, mind, and emotion such as relaxation, improved resilience to stress, neuroprotection, or analgesia. Cannabinoids come in three types: those made by plants (primarily cannabis), those made by the human body itself, and synthetic versions (human-made).

Endocannabinoids: Compounds made by the human body that interact with the same biological structures as those compounds made by cannabis producing identical or similar effects.

Cancer

There is abundant scientific literature looking at cannabis in the treatment of various cancers. Evidence suggests that cannabis is capable of direct and indirect therapeutic actions in the context of treating a number of types of cancers. The growing list includes cancers (listed alphabetically) of the bile ducts, bladder, bone, brain, breast, cervix, colon, kidney, blood, head and neck, liver, lung, lymph, mouth, pancreas, plasma, prostate, skin, thyroid, and uterus. Cholangiocarcinoma, Kaposi's sarcoma, leukemia, lymphoma, melanoma, multiple myeloma, neuroblastoma, rhabdomyosarcoma, squamous cell carcinoma–cannabis has a role to play in treating all of these and more. Here we take a closer look at cannabis and brain cancer.

There are more than a hundred different types of brain cancers. A common and malignant type is a glioma. The word *glioma* is compiled from the Greek *glia* "glue" and the suffix *-oma,* meaning "morbid growth or tumor." This type of cancer tends to form in supportive cells (glial cells) of the brain or spinal cord.

The location and size of cancerous brain tumors largely determine survivability and the various signs and symptoms likely to develop. As tumors grow, impairment increases. As a result, the whole body as well as the mind may be affected. Generally speaking, the lower in the brain structure the tumor is, the poorer the survival outcome. This is because the brain's lower portions control most vital functions such as breathing and heart rate.

Signs and symptoms cover a relatively wide range and can include altered levels of consciousness ranging from mild confusion to epileptic seizures, from odd behavior to full-blown stroke-like handicaps. Additional commonly reported symptoms include headache, visual impairments, and nausea with vomiting.

Research Highlights in Brain Cancer

A number of cases of significant improvement in patients with a very poor prognosis of glioma have been described, but the exact mechanism by which cannabinoids produce their therapeutic effects remains to be elucidated. Of the studies reviewed for this chapter all but two were limited to laboratory and animal experiments and reviews of the currently available literature, they give us somewhat limited experiences to use in making more informed decisions.

However, these pre-clinical trials do reveal numerous cannabinoids that activate a number of biological pathways which exert potential therapeutic effects. To read the review of select scientific literature, go to the Appendix.

Chemotype Considerations for Brain Cancer

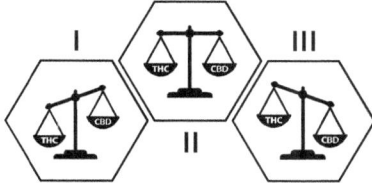

Research results demonstrated that both THC and CBD, synthetics, and endocannabinoids are potent inhibitors of brain cancer cell development and are able to induce apoptosis (cancer cell death). Each works by unique mechanisms not limited by activation of the classical pathways. Greater effects may be produced synergistically when THC and CBD are applied together. Thus all three chemotypes may be considered.

Heart Disease

The heart is a unique muscle that gets its own nourishment such as oxygen from three main vessels called the coronary arteries. In Western medicine's model, when one or more of these coronary arteries is slowly or suddenly blocked (as in the case of a blood clot) or gradually narrows (as in the case of atherosclerosis, a buildup of plaque), it becomes deprived of oxygen and the heart muscle begins to ache.

More specifically, this process is called myocardial ischemia (an inadequate oxygen supply to the muscles of the heart). Depending on severity, ischemia can lead to irregular heartbeats (arrhythmias) that when severe enough or sustained over time can produce chest pain (angina) or progress to an infarct (heart attack), a potentially life-threatening condition.

However, merely restoring blood supply to the ischemic tissue may not solve the entirety of the problem. Actual practice shows that after emergency interventions (clot-dissolving drugs or surgery, for instance) the damage often increases.

This additional trauma to an already "aching heart" is referred to as ischemia/reperfusion (I/R) injury. It is hypothesized that this damage is induced by free radicals or reactive oxygen species (ROS) such as superoxide (O_2—) or hydrogen peroxide (H_2O_2). The more ROS accumulate during a heart attack, the deeper the damage.

While healthy cells have the capacity to neutralize ROS by using enzymes—such as catalase for (H_2O_2) or superoxide dismutase for (O_2—)—that disarm or detoxify them by changing them to harmless molecules such as regular oxygen or water, already stressed or damaged cells are not able to detoxify these radicals, thus creating additional compounding events such as the production of pro-inflammatory cytokines, toxic levels of the excitatory neurotransmitter glutamate, adhesion molecules, and immune reactions where the body's defenses begin to view the damaged cells as enemies and thus begin to attack them, causing more local damage as well as systemic ones.

While cannabinoid-based therapies cannot do anything about some of the contributing factors underlying the development of heart disease such as family history, quite a few of them may also respond positively to cannabinoids, such as chronic mental-emotional stress (including their physical correlates e.g., inflammation-based stress and oxidative stress), hypertension, atherosclerosis, diabetes, and mood disorders (e.g., anxiety, depression).

Heart Disease and Cannabis: Summary of Effects

Pre-clinical trials show numerous potential mechanisms by which cannabinoids modulate cardiac function. Thus cannabinoids present new mechanisms that may at once be cardioprotective and potentially therapeutically efficacious in dealing with contributing factors as well as numerous chronic and acute cardiac events, including hypoxia (poor tissue perfusion), arrhythmias (irregular heartbeats), angina, heart attacks (myocardial infarction), and myocardial ischemia/reperfusion injury.

One of the major contributing factors to cardiovascular illness is chronic stress. Cannabis is famous for being able to calm, mitigate, and reduce the negative impact of stress on body, mind, and emotion. And, while some laboratory studies and animal experiments suggest cardioprotective effects for THC, researchers also noted as early as the seventies that THC also increases heart rates as it lowers blood pressure, thus suggesting limitations of its use in certain circumstances, for example in the presence of rapid heart rhythms (tachycardias). However, it is also hypothesized that the slight increase in heart rate might be a compensatory mechanism to adjust for the slight drop in blood pressure and thus is but a part of the body's own, continuous, and natural means to achieve adequate tissue perfusion.

In direct contrast, CBD does not produce changes in hemodynamics such as heart rate and blood pressure. To read the review of select scientific literature, go to the Appendix.

Chemotype Considerations for Heart Disease

The studies reviewed for this section have demonstrated that while both THC and CBD have been shown to possess the ability to positively reduce stress and improve mood, THC has a narrow therapeutic window, while CBD has a very wide therapeutic window. Thus, especially if you do not know your specific therapeutic amount of THC, you can avoid the potential adverse effects of tachycardia or anxiety, possibly resulting in additional cardiac stress, by choosing a CBD-rich strain (chemotype III) with no or only small amounts of THC.

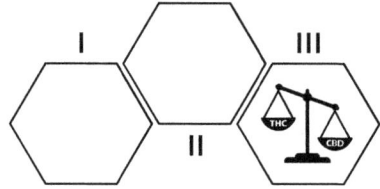

The Stigma of Cannabis

The misinformation campaign so basic to the narrative of the war on drugs has had a profound effect on our collective global psyche. Even in countries such as Holland (where medical cannabis has long been available), those who are new to cannabis often feel afraid to use it for what ails them. Fears of getting "high," fears of potential adverse effects, and fears of addiction keep many patients from what could otherwise be a remarkably safe, natural, inexpensive, and effective medicine.

The potential damaging effects of such unexamined bias and prejudice without the benefit of discernment is not just keeping people away from potential relief but may actually make things worse if not properly addressed. Case in point: Seniors represent one of the largest cohorts new to cannabis. One of the most common reasons seniors visit the doctor's office is for the treatment of various types of pain. And pain is never just a purely physical sensation. It always has a mental-emotional component to it. Studies have shown that when people feel scared, their pain threshold goes down. The opposite is also true: When people people feel safe, their pain threshold goes up. Elderly patients often arrive at the physician's office with an internal architecture of fear and worries about the "evil weed from Mexico." However, as you will see throughout this book, these prejudices and biases not grounded in science are relatively easy to overcome. Equipped with evidence-based information, patients will have increased agency about their healing process, and thus can feel empowered, and more confident, in taking the actual steps of safely working with cannabis to produce and sustain the precise effects they wish to realize.

Chronic Pain

In general, pain is defined as an unpleasant physical, mental, and emotional experience linked to real or even only expected damage to body or psyche. In mere physical terms, pain is caused by trauma from thermal forces (heat or cold), mechanical forces (pressure, tearing, cutting), chemical forces (acids, bases, toxins), or radiation (UV light, microwave, or other ionizing radiation such as X-rays). A significant number of people also suffer from ill effects of mental-emotional types of pain. Different types of pain can be divided into eight categories: acute, chronic, central, peripheral, nociceptive, neuropathic, inflammatory, pathological, and mental-emotional. Evidence shows no single treatment works on all types of pain equally well. The same is true for cannabinoid-based analgesia.

To cover all the analgesic pathways and all types of pain is beyond the scope of this book. But we will examine one particularly stubborn kind of pain that is especially resistant to standard orthodox analgesic regimens: namely, chronic (non-malignant) pain.

The word *chronic* has its origin in the Greek *khronos*, meaning "time." Chronic pain is usually considered to last more than 12 weeks and is generally less intense than acute pain.

Cannabinoids are uniquely positioned to mitigate all types of pain because cannabinoid receptors are widely spread across numerous pain-modulating pathways such as central and peripheral sensory neurons, brain regions modulating sensory discrimination, pain-regulatory circuitry in the brain stem, and affective states that regulate emotional responses to noxious stimuli. In addition, the body's own endocannabinoids may provide one of the earliest therapeutic responses to pain by unlocking one or more of the available pathways to realizing analgesic effects.

Research Highlights in Chronic (Non-Malignant) Pain

Three high-quality placebo-controlled trials conducted on humans suggest effective cannabis-induced analgesia in cases of chronic (non-malignant) types of pain. To read the entire review of select scientific literature, go to the Appendix.

Chemotype Considerations for Chronic (Non-Malignant) Pain

Taken together, these human trial results suggest that observed analgesic effects in cases of chronic pain were most likely modulated by THC, or by products containing both THC and CBD (relatively equal ratio), thus suggesting an optimal effect produced by cannabis chemotypes I and II.

Remember, these considerations are based on limited available scientific literature to

date. In fact, a number of pre-clinical studies suggest that CBD may reduce chronic pain effectively, especially if underlying inflammatory conditions are part of the chronic pain picture. Furthermore, early research results also indicate a possible synergy between CBD and opioids, potentially allowing for combined analgesic effects while reducing opioid dosages and with it the risks of adverse effects.

Diabetes

Traditionally, orthodox medicine categorizes diabetes into three kinds: Type I, Type II, and gestational. Type I diabetes was once called juvenile diabetes because it occurred mostly in children or adolescents. In Type I, the pancreas stops producing the hormone insulin. Without insulin the body cannot use the food life depends on, cellular sugar. Treatment in the orthodox paradigm consists of daily insulin injections, normally administered by the patients themselves. Orthodox medicine lacks a cure and has not yet determined the exact origins of the condition.

Research Highlights in Diabetes

Because of the well known effects of cannabinoids on fat tissue and the glucose/insulin metabolism, it is possible that there are therapeutic applications for cannabinoid-based therapies for diabetes. In addition, the known anti-inflammatory and immune-modulating properties of the plant have prompted a number of experiments relevant in the context of diabetes or diabetes-related conditions such as diabetic neuropathies. While cannabis oil has been used historically in the treatment of diabetes, and while many diabetic patients claim that cannabis lowers high blood sugar levels, and stabilizes mood changes and mental irritability, to date only one human trial has been conducted to examine the general effects of two cannabinoids, namely THCV and CBD, on diabetic patients. Of these two, THCV may provide patients with some therapeutic benefits.

Chemotype Considerations for Diabetes

Out of the 10 studies reviewed for this chapter, only chemotype I and II have been used, and thus strong trend-based chemotype suggestions are lacking. However, if we consider that THC, THCV, and CBD may show therapeutic promise in the context of diabetes, all three chemotypes may be of revelance.

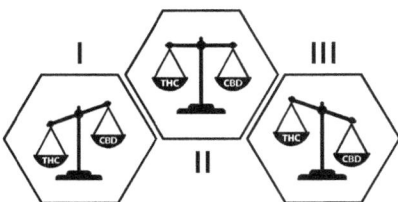

Anxiety

Generally speaking, anxiety is a normal reaction to the subjective experience of stress, such as in "performance anxiety." It can occur when anticipation of future events is associated in one's mind with thoughts and feelings not rooted in the present moment. While anxiety can be considered a normal part of life, chronic or constant anxiety can be debilitating to one's quality of life. In fact, such interference can produce very real physiological changes in the short and long term. It is estimated that almost 2 out of 10 people in the US suffer from some kind of anxiety disorder.

Western medicine considers anxiety disorders mood disorders and defines five basic types: generalized anxiety disorder (GAD), obsessive-compulsive disorder (OCD), panic disorder, post-traumatic stress disorder (PTSD), and social anxiety disorder (social phobias).

Research Highlights in Anxiety

The endocannabinoid system is clearly implicated in the experience of anxiety. Today's available literature confirms that cannabinoids, those made by the human body, synthetic versions, and cannabis constituents, modulate the biological underpinnings and mood states relevant to anxiety.

A review of over 40 studies looking at the effects of cannabinoids and other cannabis-based plant constituents included eight human trials of which a majority were double-blind, placebo-controlled trials. Results from both pre-clinical and human trials show the following trends or insights: THC and CBD activated different parts in the brain's architecture and produced opposing effects relevant to reducing anxiety. THC can produce a deep sense of relaxation and sedation, but the line between a therapeutic effect and making anxiety worse is very thin, and usage of THC ought to be carried out with caution. CBD reduces anxiety but does not produce sedative effects. CBD produces mood improvements that can be harnessesed in the treatment of anxiety and other mood disorders such as depression.

Chemotype Considerations for Anxiety

These data indicate that the amount of THC, the primary psychoactive cannabinoid of cannabis, is a source of major concern and focus in determining a safe and optimal dose. A chemotype I with more THC than CBD has indeed a very narrow therapeutic window where the line between a therapeutic effect is indeed razor thin (you may remember the example of Maureen Dowd from the introduction). Even a chemotype II with a balanced THC and CBD ratio and a medium-size therapeutic window may be still be too much for some sensitive patients. A chemotype III (with very low to trace amounts of THC%) may provide the safest option for people wishing to avoid changes in cognition. For a more complete review of the literature, go to the Appendix: The Science behind the Top 10.

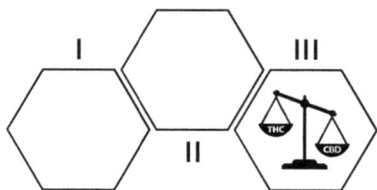

How to Approach a Physician, Nurse, or Other Health Care Provider with Questions

Because cannabis has long been stigmatized, many people are understandably nervous and unsure of how to bring up the topic and ask the right questions.

Here are a few things to consider. We are truly living in a new era where the culture of medicine has shifted from a top-down model of physician control over all aspects of patient care to a more collaborative model in which decisions are made together by patients and their health care providers considering the evidence, their experience, and the patient's preferences alike. Additionally, most health care providers are fully aware of the limitations of a pharma-only approach. Most of them wish they had better and safer healing tools at their disposal. This is especially true for those specialties that deal with chronic conditions for which there is no cure within the orthodox system of Western medicine but for which cannabis may show promise.

Also, please consider that your health care providers are humans too and thus may also have been misinformed by the narrative surrounding the war on drugs. Additionally, education in medical school did not include anything about the cannabinoid health sciences (even now, only a few medical schools offer any kind of introduction to this field). So for many of them this particular path to health and healing is simply new territory. However, it is certainly not the first time a health care provider has been challenged or motivated by the presence of new information, as evident in the steady increase of alternative, complementary, and integrative approaches has been steadily rising over the past twenty years. And, usually after a phase of initial growing pain, many of them are now embraced and effectively utilized. This is also certainly true for cannabis-based medicine, as all states that have medical programs clearly show. So, in a way, the map has been made. Here are a few questions you might find helpful in approaching your physician or nurse about working with cannabis for your symptoms or condition.

- Do you have any experience in using cannabis as medicine for what ails me? If not, would you be willing to give me a referral to a cannabis-experienced health care provider?
- I am considering using cannabis for my condition. Would you be willing to work with me to dial in the optimal plant constituents, forms, and dosage?
- Would you help monitor my progress if any of the drugs you are prescribing for me could be reduced or eliminated as a result of concomitant cannabis use?
- Do you think that any of my pharmaceutical drugs potential adverse effects could be diminished by concomitant cannabis use?
- Do you think that any of my pharmaceutical drugs' potential adverse effects could be worsened by concomitant cannabis use?
- If you do not know the answer to my question, could you find out or do you know somebody who could?
- Would you like me to share some information I found valuable?

Neurological Conditions

Orthodox medicine considers multiple sclerosis (MS) a chronic, inflammatory, and degenerative neurological illness with no cure and no known cause. In fact, MS is one of the most common neurological diseases.

Multiple sclerosis is characterized by the breakdown of some of the thin sheets that cover the brain and spinal cord. These fat-based myelin sheets normally provide insulation and protection, but when lesions occur, nerve impulses misfire across the broken insulation, causing a variety of debilitating symptoms.

Research Highlights in Multiple Sclerosis

The results of the 34 studies reviewed for this section share a common denominator. For patients suffering from MS, cannabis may be a potent ally in alleviating the disease's symptoms. Furthermore, cannabinoids may slow the progression of the illness itself and thus provide an improvement to long-term survival and quality of life.

Most clinical studies report adverse effects of cannabis use in addition to benefits dependent on dose and form. Adverse effects include reduced balance and posture, nausea, and dizziness; and at high dosages, negative psychological symptoms such as anxiety. Results also indicated that when used within the proper subjective therapeutic dose, cannabinoids' potential adverse effects are usually well tolerated and negligible, especially when compared to the beneficial effects.

Chemotype Considerations for Multiple Sclerosis

Of the studies reviewed, 7 used a chemotype I, 16 a chemotype II (mostly in the form of Sativex), and 2 used a chemotype III. The emerging trend here is the use of a chemotype II.

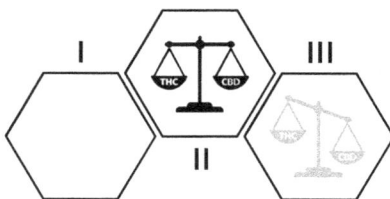

Sativex is a whole-plant extract that contains relatively equal amounts of THC and CBD (2.7mg THC:2.5mg of CBD). Due to the abundance of clinical trials that have tested Sativex in the treatment of primary and secondary symptoms of MS, many physicians new to cannabis-based prescribing feel more comfortable recommending the pharmaceutical version. However, for those of you who wish to use cannabis flowers or other forms of cannabis containing products made from equal amounts of THC and CBD, you will find them in most dispensaries that tend to focus on serving patients. For a more complete review of the literature, go to the Appendix: The Science Behind the Top 10.

Insomnia

Insomnia is difficulty or inability with falling or staying asleep. Poor quality of sleep resulting in waking up tired or exhausted is a common symptom of people suffering from insomnia. It is estimated that in the United States alone, a third of all adults deal with insomnia at some point in their lives. Women are suffering from insomnia at twice the rate as men, and over half of all seniors over 60 suffer from this condition.

Western medicine divides insomnia into primary and secondary. In primary insomnia, insomnia is the sole problem. Secondary insomnia is produced by other causes. Insomnia is further divided into acute and chronic. Acute insomnia is considered temporary, short-term, and self-correcting—new work hours, jet lag, or significant time-zone differences. Chronic insomnia has an ongoing and underlying problem—numerous diseases, pharmaceutical or recreational drug use, or pain, for example.

Research Highlights in Insomnia

The results of a review of 25 insomnia and cannabinoid studies provide the current evidence to help us make more informed decisions in relation to insomnia. Effects of using individual cannabinoids are dose-dependent. Too much or too little of a single cannabinoid can have unwanted effects in general and in time specifically, as you may find initial relief but a couple of hours later your symptoms are getting worse. The results also show that two of the prime cannabinoids, THC and CBD, can create a synergy. It seems that both taken together, or even better as a full-spectrum product, balance each other to maximize the most desired effects: greater ease of falling asleep, deep sleep, and no residual drowsiness upon morning rising. To read the review of the select scientific literature, go to the Appendix.

Chemotype Considerations for Insomnia

Creating the best possible sleep synergy by discovering the best chemotype for insomnia depends on your own needs, which differ significantly from one patient to another. However, with a bit of attention to detail, it may be easier than you think.

We know that THC promotes sleep and reduces the time it takes to fall asleep, but can also cause morning drowsiness; at higher dosages, especially for susceptible patients, it can cause panic and anxiety, thus making insomnia worse. CBD at higher concentration also supports deep sleep but at lower concentrations induces wakefulness and thus balances out the sedating qualities of THC.

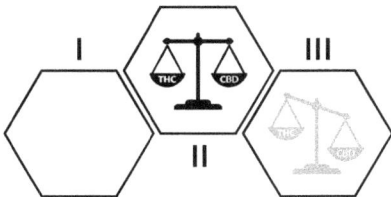

Gastrointestinal Conditions—Colitis

Colitis, also commonly referred to as inflammatory bowel disease (IBD), simply means inflammation of the large intestine. There are a number of types of colitis which are identified by their underlying causes, such as autoimmune colitis, ischemic colitis, allergic colitis, microscopic colitis (i.e., collagenous colitis, lymphocytic colitis), ulcerative colitis, chemotherapy-induced colitis, infectious colitis (e.g., salmonella), or idiopathic colitis (of unknown origins). To date, there is no cure within the model of modern medicine, and treatment is limited to supportive therapies such as hydration via IV fluids, pharmaceutical intervention (e.g., steroids to depress the immune system), iron supplements to balance anemia, or surgery. Some progress has been reported with certain dietary regimes and microbiome supportive supplements. Colitis is different from irritable bowel syndrome (IBS), a chronic and typically non-inflammatory syndrome of the large intestine.

Research Highlights in Colitis

Case reports from cannabis-using IBS patients suggest that cannabis may be effective in managing some symptoms, especially nausea, diarrhea, stress, cramps, weight loss, and lack of appetite. A placebo-controlled clinical trial was conducted on 21 patients with significant manifestations of Crohn's disease who did not respond to the usual pharmaceutical regimen consisting of steroids, immunomodulators, or anti-tumor necrosis factor-α agents. Results showed that inhaling from a cannabis cigarette (containing 115 mg of THC) twice daily produced complete remission in 5 of 11 patients in the cannabis group and 1 of 10 in the placebo group. Three patients in the cannabis group were weaned from steroid dependency.

Chemo-type Based Specific Considerations for Colitis

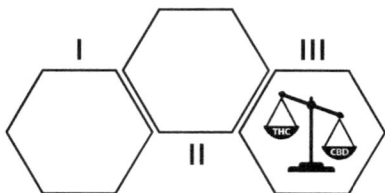

Of the studies reviewed for the colitis section, CBD emerged as the primary cannabinoid followed by THC. As such, all cannabis chemotypes may be efficacious in the treatment context of colitis. For a more comprehensive review of the literature, go to the Appendix: The Science Behind the Top 10.

The Difficulty of Finding Research That Is Relevant for Me and My Condition

Over the past decades, research into cannabis and its therapeutic potential has grown almost exponentially. However, obtaining the practical and relevant information that can guide you or any other particular patient to make more informed decisions can be a challenging and arduous process. Even if you are willing to spend the time and energy to go to the largest repository of published scientific papers at the US National Library of Medicine compiled by the National Institutes of Health, you are faced with the Herculean task of weeding irrelevant entries from those that could be meaningful.

Consider an example request: Suppose you want to find out what the currently available literature says about using cannabis for chemotherapy-induced nausea and vomiting (CINV). If you punch in the keywords chemotherapy-induced *nausea* and *vomiting* and *marijuana*, you will get over 120 entries covering studies from 1975 through May 2020. Now your task is to read through each of their respective summaries and pull out the pieces of information that make a practical difference to you. What is the current evidence-based consensus? Does it say cannabis helps to mitigate CINV or not? And if the results indicate actual therapeutic evidence, what is that consensus's level of confidence? Which type of cannabis, which specific plant-constituents, form, dose, is thought to work best? What type of cannabis is considered safest, and which ones present with the potential of adverse effects?

To make this challenge a lot less daunting is the purpose of this book and the material laid out in the appendix. For instance, in the section "the Science Behind the Top Ten" the book takes you through ten examples of how to line up the currently available data in a practical manner to maximize your ability to discern what works and what doesn't.

Furthermore, you may want to explore two other books I have authored that align, complement, and expand on the material suggested here: *The Cannabis Health Index* and *Breaking the Cycle of Opioid Addiction: Supplement Your Pain Management with Cannabis.*

Cough

Coughing is perhaps the most common affliction of humanity. It remains one of the most vexing problems for patients and physicians alike. To date, there is no cure for chronic coughs (or colds or flu) within the orthodox model of medicine.

For the past 200 years of modern medicine, opiates such as morphine or codeine have been used to treat a cough with some degree of relief. However, especially in light of the recent opioid epidemic and the possibility of addiction or abuse, and commonly reported adverse effects such as constipation, fewer physicians feel inclined to use them.

Over-the-counter drugs containing dextromethorphan such as TheraFlu, Vicks, NyQuil, or Robitussin are commonly used with some degree of success, but a review of available studies conducted on the efficacy of dextromethorphan showed difficulty in producing clinically significant effects. There are a number of other OTC drugs such as Actifed or Sudafed that may suppress a cough reflex. However, they also contain speed-like substances (amphetamines) that can change your emotional state and make you anxious, jittery, and keep you from sleeping, as well as raise your blood pressure and your heart rate.

Similar to a common cold or a flu, the primary cause for an infection-based cough is a viral infection. Antibiotics target bacteria only and are ineffective in viral infections, as well as destroying beneficial intestinal flora. Single-ingredient antihistamines may provide some minor relief in cases of allergy-based cough. The need for relief from coughing remains largely unmet.

A word about coughing: A time-limited productive cough (bringing up material) is the lungs' way to clean themselves, a beneficial and necessary function. A chronic cough is an indication of an underlying condition (e.g., COPD, emphysema) or a maintaining cause (e.g., infection or inflammation due to allergens, toxins, or microbes; lowered immunity) and should be evaluated by a physician.

Warning: While most coughs will simply run their course and are by nature self-limiting, certain complications or conditions may warrant urgent treatment, especially if symptoms include shortness of breath, chest pain, cough-induced insomnia, fainting, vomiting, incontinence, hernias, nausea or vomiting, pale and sweaty skin, prolonged or very high fever, or tissue damage of the rib cage.

Cannabis and Cough: Summary of Effects

Cannabis has time-proven anti-inflammatory, antispasmodic, and bronchodilating properties, all of which may play a part in producing the therapeutic impact of the herb as a potential anti-tussive (a drug used to relieve coughing).

Chemotype Considerations for Cough

The studies reviewed for this summary suggest that specific cannabinoids (primarily THC and CBD) and the body's own anandamide were involved in the amelioration of coughing albeit achieving results by different mechanisms. Thus all three chemotypes may provide relief. To read the review of the select scientific literature, go to the Appendix.

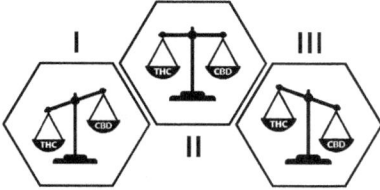

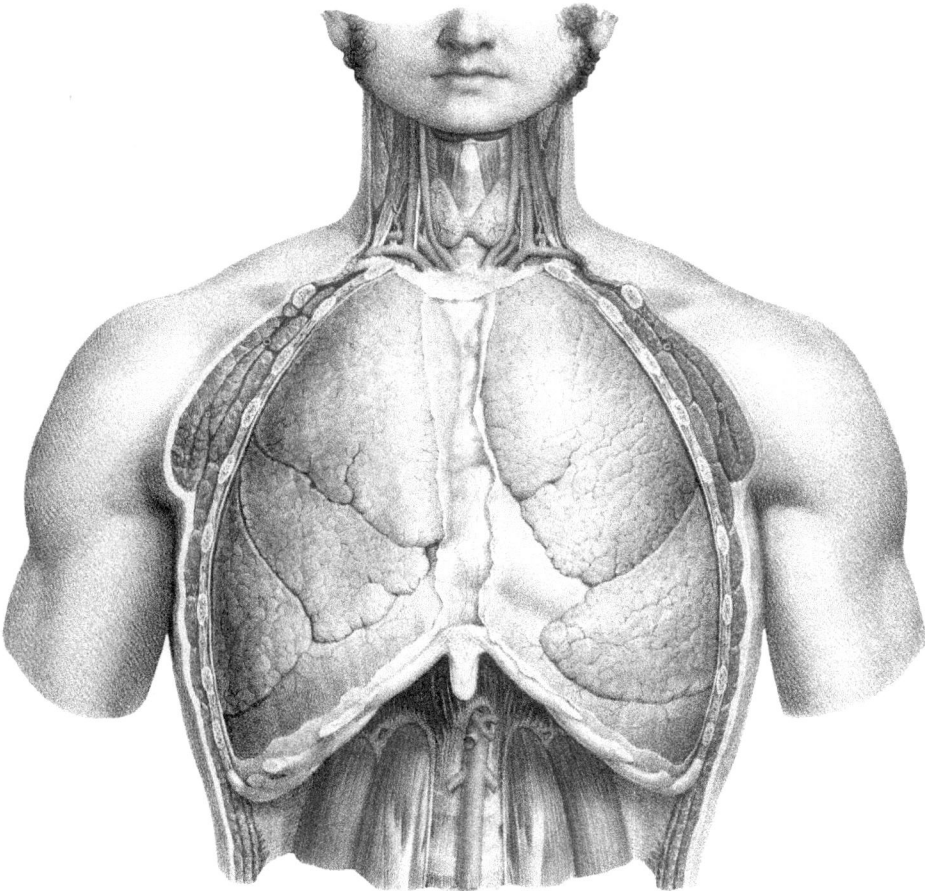

Skin Conditions

The number one reason people go to the doctor is help in treating many different skin conditions. Evidence has shown that cannabis may provide relief with acne, aging skin, dermatitis (eczema), skin cancers, itching, psoriasis, scleroderma, and seborrhea. To look at all of them is beyond the scope of this book, so one common condition, acne, will be discussed here.

One of the most prolific skin conditions, acne (like most chronic illnesses) has multiple underlying pathologies such as stress, poor diet, hormonal imbalances (testosterone tends to increase and estrogen decrease sebum production), inflammation, genetic predisposition, overproduction of sebum (oily secretion), and excessive proliferation of sebocytes (cells that produce sebaceous glands). Orthodox medicine has no cure, and effective treatment is wanting.

Sebum is produced by sebaceous glands, which are unevenly embedded throughout the dermis (middle) layer of the skin. As people who deal with acne know all too well, sebaceous glands tend to be highly concentrated in the face, scalp, and back (where acne tends to proliferate), while there are none or but a few of them in the soles of your feet or in the palms of your hand. Hair follicles also pass through sebaceous glands, and thus overactive glands contribute to greasy hair and scalp.

When balanced, however, sebum functions to protect the skin and hair. Sebum waterproofs the skin and protects from the loss of moisture by providing a thin layer of oily substance. Additionally, sebum is slightly acidic and thus capable of neutralizing many environmental toxins (which tend to be alkaline in nature); it also has anti-microbial properties and thus represents a first line of defense against certain fungi, viruses, and bacteria.

Depending on where the cells are located in the layers of the skin, excess sebum can lead to whiteheads (beneath the skin) and blackheads (on the surface of the skin) as well as pimples, zits, or pustules commonly referred to as acne.

Too little sebum and you have dry, itchy, or cracked skin, while too much may lead to the clogging of sebum and dead skin cells, which increases cellular pressure and may cause rupture—releasing sebum and other content into the surrounding tissue, causing acne. Patients with acne may also be at risk for secondary bacterial infections, which may prolong the healing process.

Skin Conditions and Cannabis: Summary of Effects

Cannabinoids interact with a number of receptor sites in the human skin and thus may present multiple new and simultaneous targets for the treatment of acne. These naturally occurring cell receptors have been shown to balance skin growth and sebum production, and induce systemic and localized anti-inflammatory effects, all of which play a role in acne.

THC has been shown to increase sebum production, and so current research would suggest not using a THC-rich strain. Patients who need more THC for other reasons such as stress management (relaxation) often consider using a chemotype II with a nearly balanced THC:CBD ratio and a relatively low concentration of THC. For a more complete review of the literature, go to the Appendix: The Science Behind the Top 10.

Chemotype Considerations for Skin Conditions

Data suggest that CBD has a positive localized effect in the skin where acne develops and thus seems uniquely positioned to address a number of the underlying causative pathologies of acne. More specifically, CBD's combined lipostatic, sebum anti-proliferative, and general skin-specific anti-inflammatory effect has potential as a therapeutic agent for the treatment and prevention of acne. In addition, CBD has known therapeutic abilities in reducing certain moods commonly associated with acne such as anxiety or depression. CBD confers mood improvements without euphoria or changes in cognitive function. Thus research leans toward considering a chemotype III (high in CBD, low or negligible THC) for the prevention or treatment of acne and its commonly associated negative affect (emotions).

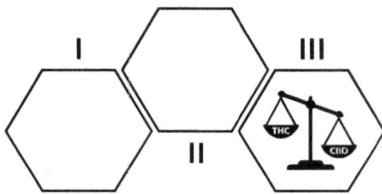

These are some examples of how using the currently available scientific evidence allows you to identify the most appropriate cannabis chemotype and to make practical and more informed decisions about your health care. Choosing the most appropriate cannabis chemotype is only the beginning of achieving and sustaining the therapeutic effects you desire or need. Before we move on to the practical steps for finding the type of cannabis that is optimal for you, it helps to look at some of the confusions that make this process more complicated than it needs to be.

Key Points, Chapter 1: Top 10 Reasons People Seek Medical Care, and the Appropriate Type of Cannabis for Each

The currently available scientific literature highlights these cannabis chemotypes for the following:

- Cancer I, II, and III
- Heart Disease III
- Chronic Pain I and II
- Diabetes I, II, and III
- Anxiety III
- Neurological Conditions II
- Insomnia I and II
- Colitis
- Cough III
- Skin Conditions III

Chapter 2

Cannabis as a Healing Plant: Confusion and Insight

In this chapter you will learn the following:
- How changes in research technology created their own set of problems and how to avoid them
- How to understand the many names of cannabis
- How to go beyond the sativa vs indica distinction
- What the biphasic effect is
- What the safety parameters of cannabis are: death, adverse effects, and addiction potential
- Which families of cannabinoids tend to occur in high-enough concentrations to make a difference

---

The cannabis landscapes for patients, their caregivers, and their health care providers are far from optimal. The war on drugs is a failed policy; the Western medical system is not designed to easily integrate cannabis as medicine. Simply put, the way the plant has been classified and studied, or not studied, is not ideal. Despite these roadblocks, people have not been deterred—more and more people are discovering that cannabis is an effective medicine for a wide range of ailments.

Before we dive in and detail how cannabis produces therapeutic effects—and more important, how to produce and sustain the precise effects you or your patients need—here are some of the issues that have made the situation confusing.

These issues include decades' worth of research materials and their deficiencies (e.g., compatibilities); changing general naming conventions; the indica versus sativa distinction falling apart; and the speed with which emerging concepts about cannabis healing are moving front and center in the cannabinoid health sciences.

Research and Its Drawbacks

Much of today's cannabis confusion is rooted in outdated knowledge. Early 19th-century chemists such as Wood, Spivey, and Easterfield, and Raphael Mechoulam and Yehiel Gaoni in the 20th century, were true pioneers in the field who discovered the first scientific puzzle pieces that eventually gave birth to a new field in medicine. Many of their findings are still valid and informative today. However, limited understanding of cannabinoid chemistry, old technology to detect and analyze their findings, and quality of cannabis place some of the early discoveries in apparent opposition to newer ones.

For instance, many past studies were conducted using "seized" cannabis that differed significantly in quality, origin, and age, and thus in presence and concentration of actual plant constituents. Cannabis until the '70s contained relatively low percentages by weight of THC. Since then, thousands of new cannabis hybrids have been bred, most with the aim of increasing THC concentration. Studies using cannabis with low versus high THC concentrations can produce different outcomes.

Over time, however, and as more and more studies become available, a cluster of trends begins to emerge. For instance, when five or more studies signaled similar insights, that trend would statistically stay most often and be eventually confirmed. While conditions with five or fewer studies may still indicate the truth about a matter, but more studies would significantly indicate probable or even actual evidence that supports the use of cannabinoids for that particular ailment.

Another example of confusion awaiting clarification is the vague official scientific designation of individual cannabinoids (the nomenclature), because the many constituents of cannabis are still undergoing precision categorization. Regular updates, reviews, and proposals are beginning to address these challenges of defining the basic cannabinoid inventory and their pharmacological impacts on the body. A single naming system would solve this problem, but for now there are still five numbering and naming systems in use.[1] As a result, you may see different studies referring to the same primary psychoactive compound with different representations such as $\Delta1$-THC, $\Delta3$-THC, $\Delta6a$-THC, and $\Delta9$-THC. Attempts to address these perplexing naming issues are in progress.

Other Resources to Explore

Readers who wish to deepen their own learning may want to consider three research platforms that anyone can access and use for free: **PubMed, PubMed Central**, and **Cochrane Library**. Here is a brief review of their strengths and limitations.

PubMed is an online portal produced and maintained by the United States National Library of Medicine that offers access to the Medical Analysis and Retrieval System, aka MEDLINE, which stores the abstracts and a limited number of full the text of scientific research papers published published between 1951 and the present. The strengths of PubMed are also its weakness (it casts a wide net). In other words, your keyword searches can yield an overabundance of results, which may contain the specific information you are looking for but also often include an overwhelming amount of studies, of which many may have nothing to do with your keyword searches. Another weakness is that many searches yield only the abstract of the studies you are looking for and thus make it difficult to find the important details of research results and findings. https://www.ncbi.nlm.nih.gov/pubmed/

PubMed Central (PMC) is also an online portal to a vast repository of scientific stud-

ies. However, in contrast to PubMed, it contains access to the full text of all its study articles, making deep learning a real possibility. Its limitations are twofold. One, it contains relatively new entries, from 2000 to the present; two, it produces significantly more false positives than PubMed, thus making some research endeavors very time-consuming affairs.
https://www.ncbi.nlm.nih.gov/pmc/

The Cochrane Library is set apart from PubMed and PMC in that it specializes in evidence-based repositories, systematic reviews, and meta-analyses to arrive at the highest quality of discerning data available, making it a great resource if you are looking to determine and apply actionable efficacy data. The drawback is that due to its very high standards of inclusion criteria, a lot of otherwise actionable and helpful data will not show up in its trends, results, and data reporting. Also, be advised that older but still valuable studies from before 1996 will most likely not be included, further excluding otherwise potentially useful data.
https://www.cochrane.org

What's in a Name? Classifying Cannabis

The different names for cannabis can be confusing. To be clear, there is only one genus (biological classification or rank) of cannabis, which is divided into three primary species: **sativa**, **indica**, and **ruderalis**.

Another commonly heard term is **strains**. In a way you can think of a strain as an attempt to further differentiate, or to create an additional value ranking, within one of the three species of cannabis. When you see a "sativa strain" reference, or someone mentions "indica strain qualities" or a "hybrid sative-indica strain," people are usually trying to ascribe particular qualities to a certain plant within a species of cannabis of which there are literally thousands.

Genotype and **phenotype** further differentiate the plant. The plant's genetic composition, or DNA blueprint of life, is referred to as a genotype. It contains all the genetic variations or possibilities of expression. Phenotype describes the way a specific environmental signal activates a set of genes while leaving others inactive—how the plant physically expresses itself (what it looks like and does). You may find it helpful to think of phenotypes as siblings sharing some of the same visible traits but not others. In fact, strain, hybrid, and phenotype are often used interchangeably without much confusion. In this book I will try to be clear about their specific usages, but but no need to fret if you mix them up.

A **hybrid** is produced by crossbreeding the qualities of two distinct species—more specifically, most cannabis hybrids are crosses of sativa and indica. You may find it helpful to think of hybrids as offspring from a marriage between an indica and a sativa. And, as any offspring would, they carry a combination of traits from their parent species.

Additionally, there is the name **hemp** (aka **industrial hemp**), which is generally used to describe industrial crops raised for the plant's fiber (to be used as rope, textiles, paper, for example) or nutrient content (e.g., hemp seed milk, protein, fat). While hemp is usually devoid of or contains only tiny amounts of cannabinoids (depending on jurisdiction, the THC content must usually be less than ~0.3% THC for a plant to be classified as hemp), it is also commonly used as a source for harvesting CBD. Hemp is also becoming popular in the making of building materials (e.g., hempcrete, biofuel, plastic).

And, of course, there is the name **marijuana,** which has its origin as a slang term in Mexican Spanish and was used to denigrate the plant and its users by those who launched the US "war on drugs" decades ago. Still, it is a term commonly used and searched for, when researching cannabis.

You may have come across the word **sinsemilla**, which is derived from the Spanish *sin* (without) and *semilla* (seed). It refers to a growing technique whereby any male pollen is kept away from female plants. Without pollen, the female plant produces no seeds but large, dense flowers, abundant in cannabinoids.

And, lastly there is **landrace**. This term refers to a plant that is indigenous (a native part of a specific region) in an environment so unique that the species develops its own characteristics such as a specific look, ability, resilience, or distinctive genetic diversity.[2] Landrace cannabis plants tend to be unique in form, smell, taste, and of course the effects they produce when the plant is inhaled or ingested. Molecular biologists will track a landrace to better understand the spread of agriculture in prehistory, to maintain biodiversity, and to explore historical origins and learn from indigenous uses.[3]

A landrace's characteristic genome, or "fingerprint" of DNA, emerged over hundreds or thousands of years, forged in the particular soil, weather, and light of a specific geographic region—for example, the foothills of the mountain range known as the Hindu Kush, an extension of the Himalayas. A landrace may also be recognized as an original type of cannabis growing in harmony with nature and all of her spirits—those conjured by shamans of old as well as native creatures that appreciate its culinary delights, such as the local grasshoppers.

Landraces are physical vestiges of humans' oldest connections with the cannabis plant. They are sought after by growers and connoisseurs alike because they tend to be resilient, full of vital force and strong genetic materials, often with novel cannabinoid and terpene profiles. Landraces tend to be prolific and produce an abundant harvest of seeds and flowers, making them attractive to breeders aiming to create new hybrids or to highlight specific qualities and effects.

When a new landrace is discovered, the "hunter" brings his find (and prized possession) to the laboratory, where its species identification, terpene content, and, most important, cannabinoid-based profiles are revealed. Remember the key information is centered on three numbers: the amount of THC, the amount of CBD, and the ratio between them. In other words, the cannabis chemotype is identified.

A Bit More on Strains

You may have wondered why there are so many different strains and what the reasons for these variations are, or why many of them are subject to unorthodox naming conventions (e.g., Girl Scout Cookies)?

There are a number of factors that determine this vast abundance in varieties of cannabis strains, some of which include the type and quality of genetic source materials (e.g., seeds, clones), the kinds of environmental signals that interact with the plant's receptors such as quality of light (e.g., sunlight, bulbs), water (e.g., mineral content), air (e.g., filtered or not), temperature, nutrients, the presence or lack of beneficial/unhealthy soil bacteria, the presence or lack of pests, the use of pesticides and what kind, organic versus synthetic growing practices, indoor versus outdoor, weather; and let's not forget the ineffable possibility of the plant's mood (a grower friend of mine swears by playing classical music to increase her yield).

As mentioned earlier in this chapter, there are literally thousands of strains whose names simply reflect the breeder's choice, while some are derived from the "urban legend" type of cannabis folklore where perhaps commonly shared desires or even experiences crystallize to line up a quality with a descriptive name. Here are a few of the more creative or funny examples: CannaSutra, Train Wreck, Obama Kush, Skywalker OG, Purple Monkey Balls, Tiger Blood, Schnazzleberry, Purple Urcle, Gorilla Panic, African Queen, Charlie Sheen, Yellow Amnesia, Alaskan Thunderf…, Dopium, Pink Panty Dropper, Laughing Grass.

The Sativa vs. Indica Distinction Is Falling Apart (This Is a Good Thing)

Patients need to know how to achieve desired therapeutic effects and how to sustain them. Much of the current industry-wide system for choosing cannabis or cannabis-based products is still based on the assumed therapeutic qualities of cannabis's main species, sativa and indica, a distinction that has informed choices for decades. Yet it is woefully imprecise.

Skinny leaf **Sativa** is considered:	Thick leaf **Indica** is considered:
Generally stimulating, energizing, uplifting	Generally sedating, relaxing, grounding
Generally, more mental/emotional	Generally, more physical
Usually more extrovert	Usually more introvert
Best for daytime use	Best for nighttime use
Increases alertness	Sleeping aid
Considered with depression	Considered with anxiety
Pain relief, muscle relaxant	Pain relief, muscle relaxant

These assumed medical properties of sativa and indica are not evidence-based and thus are medically imprecise, confusing, and not consistent in producing therapeutic effects.

Consider this: Two people use identical seeds or clones to grow a specific cannabis species—but even subtle differences in their growing environments can significantly change the chemical composition of all plant constituents. Environmental (epigenetic) signals such as temperature, nutrition, and quality of light will initiate specific changes in genetic expressions without changing the plant's genome or DNA. Thus, even though the two cannabis species are genetically identical, the therapeutic (or adverse) effects they engender could be as different as mixing up a Zoloft with a Vicodin. Furthermore, none of the assumed qualities are evidence-based, making any success you may have had a good guess at best, further complicating consistently repeatable results. (The ability to repeat results is a hallmark of effective therapy.) While using the old system may still have value in the general context of breeding or as a way to discuss terpene profiles, for example, relying on it in a medical context also makes it nearly impossible to use the abundance of available scientific literature to align types of cannabis with the specific patient populations that would benefit from them most.

Guessing what may work for a specific patient based on sativa versus indica distinctions is the prime reason for the adverse effects still experienced by too many people. This is true for newcomers as well as for experienced users. Take a look at Donna's story.

Donna's Story

Donna, a baby boomer, was born and raised near the stretch of green that is the panhandle to Golden Gate Park, in the heart of San Francisco. As a child of the '60s, young-adult Donna was no stranger to inhaling, but it has been decades since she used cannabis. About five years ago she began suffering from a debilitating anxiety and engaged her physician for treatment and care. After trying several pharmaceutical drugs, she grew disenchanted with the significant adverse effects, high costs (she did not have insurance), and the idea of managing her condition solely with prescription drugs for the rest of her life.

As fate would have it, Donna ran into an old friend named Ella with whom she had shared her early marijuana experiences when they lived together in an old redwood-walled Victorian flat in the Haight-Ashbury neighborhood. Ella said she too had problems with anxiety, but was now free from its stranglehold thanks to the "mindful use of cannabis," as she put it. Over a few months, working with cannabis helped Ella not only to safely reduce her anxiety symptoms but also to discover underlying stressors and contributing factors. She pulled up the roots of her anxiety and freed herself from chronic stress, fear, and worries—without a major hit to her wallet.

Inspired, Donna obtained a prescription for cannabis from her doctor and went to her local dispensary in Berkeley, California. The budtender, relying on the assumed properties of sativa versus indica, suggested an indica called Granddaddy Purple for her anxiety.

Donna went home and inhaled. She did not notice any changes, so after a couple of minutes she inhaled some more. To her surprise, her anxiety symptoms suddenly increased and progressed to a full-blown panic attack. An already chronically difficult condition became a horrifying and acute event. What went wrong?

At the dispensary she shared her experience. The budtender suggested that, most likely, Donna had been impatient—instead of going slowly and carefully monitoring her emerging psychological and physical changes, she jumped ahead and inhaled too much too quickly.

Donna returned home committed to being more patient. She went slower and used less. But again she found herself feeling more anxious than normal, about 10 minutes after a couple of inhalations. Cannabis apparently helped Ella's suffering from the same illness— what was different? Donna began reading up on the subject.

She discovered that THC can produce a dose-dependent and concentration-dependent increase in anxiety in vulnerable individuals—but another cannabinoid, CBD, might tame THC's psychotropic effects. And that could produce a more therapeutically balanced effect for her anxiety.

Equipped with new evidence, Donna asked the dispensary budtender if they had laboratory test information on her last purchase, Granddaddy Purple. Indeed they did. To her surprise, the THC (21%) to CBD (0.1%) ratio was nearly 210 to 1. This time Donna selected a strain called Blue Widow with a lower THC (10%) and a higher CBD (5%) content, a ratio of 2:1.

She went home and inhaled slowly, making sure to keep track of her sensations and mood. After three medium-size inhalations, she felt her symptoms subside and her mood improve. For the first time in what seemed like a lifetime, Donna experienced what it felt like to be happy, to feel safe, and to be content. She thought she had found her ideal medicinal strain. A week later Donna refilled her prescription with more Blue Widow. She repeated her formula, but the anxiety got worse again. What was happening? It worked so well the last time!

Donna returned to check the test results. It turned out that the dispensary used a website and app that published only ranges of tested results, pooled from a number of growers and laboratories. Thus the ratios and concentrations for Blue Widow varied within a given range. Her latest Blue Widow purchase contained a nearly 5 to 1 ratio, upshifting the THC content to nearly 20% while downshifting the CBD to about 4%—more than enough to account for her uneven results. While it is great we have testing today, ranges offer only a very limited basis to discern what is optimal for you. Try to buy only clearly tested and marked cannabis flower or cannabis-containing product to ensure sustainable results.

Biphasic Effects: Same Medicine, Different Result

Cannabis and isolated cannabinoids can have opposite effects on the individual patient. For instance, a relatively low dose of THC can reduce anxiety, while a higher dose can cause anxiety or even panic attacks. In contrast, and perhaps counterintuitively in some children a high dose of CBD can cause overstimulation, especially in patients with ADHD or autism, while at lower dosages CBD can produce a calming effect. Research has shown that the likely basis for cannabinoid-induced biphasic effects rests on our body's modulation of the neurotransmitters GABAB and glutamate.[4] The very real possibility of biphasic effects underlines the importance of starting slow, using only gradual increases in dosages, careful monitoring, and tracking your specific response.

Later on in chapter 5, the CannaKeys Process will take you through all the actionable variables and teach you how to refine this process in order to arrive at a suitable prescription for yourself.

Can Cannabis Kill You? Adverse Effects and Addiction

As early as 1973 researchers tested the lethal dose of oral THC on rats. It took as much as 800–1900 mg/kg to kill the animals.[5] Using body weight as the sole criterion, this study suggests that 200 grams of herb per kilogram of body weight would be required to approach a lethal dose in humans. Accordingly, a person weighing 70 kg (154 lbs) would need to consume 14 kg (31 lbs) of herb to approach a fatal dose. A 2004 review conducted by physicians from various academic institutions and hospitals was much more conservative—it concluded that "628kg of cannabis would have to be smoked in 15 min. to induce a lethal effect."[6]

> *Because cannabinoid receptors, unlike opioid receptors, are not located in the brainstem areas controlling respiration, lethal overdoses from Cannabis and cannabinoids do not occur.*[7-11]

> —National Cancer Institute

Adverse effects may include an increased appetite, reduced attention span, red sclera (reddening of the normally white part of the eyes), dry mouth, and decreased cognitive and motor skills. Other adverse effects, more common when ingested or when used at higher than your subjective therapeutic dose, include ataxia (unsteady gait), aphasia (inability to speak clearly), unusual perceptions of all senses including hallucinations, anxiety (addressable through reassurance), a slight increase in heart rate, which is thought to be a compensatory mechanism to account for a subtle drop in blood pressure, and panic, especially upon first-ever use (moderated by reassurance).

Some observational studies have concluded that consuming cannabis as an adolescent may slightly increase one's risk of developing schizophrenia later in life. While cannabis is not itself considered a causal factor for schizophrenia, in some instances it may be a co-factor. Based on the current evidence, it would be prudent for adolescents or young adults with a known family history of psychosis or schizophrenia to stay away from cannabis or any other mind-altering substances, especially speed-based drugs such as cocaine and methamphetamines, which tend to have the highest liability potential.

Does the use of cannabis lead to addiction? The fifth edition of the Diagnostic and Statistical Manual of Mental Disorders (2013) now recognizes the diagnosis of Cannabis Use Disorder, which includes signs and/or symptoms such as craving, interference with social obligations, development of intolerance, and withdrawal (irritability, anxiety, insomnia, lack of appetite, dysphoria, for example). While the majority of cannabis withdrawals are symptoms in psychological terms, the DSM-V also includes possible physical symptoms (one or more of headache, abdominal pain, sweating, fever, chills, or shakiness) that may occur after a person stops using cannabis after prolonged (a few months) and heavy use of the plant.

Determining the addictiveness of a substance in terms of actual and precise numbers is a difficult process. Due to the lack of comparative human trials and a number of variables (e.g., physical, psychological, and cultural), the answer to which substance is the most addictive remains quantitatively elusive as well. However, it may be helpful to compare the addiction potential of cannabis to other substances.

Consider the following three research results.

Trial 1 (2015): "The addictive potential and the seriousness of adverse effects associated with these drugs (cannabis and synthetic cannabinoids) is less than those associated with some controlled but legal drugs, including prescription opioids, alcohol, and nicotine."[12]

Trial 2 (1997) showed that the most addictive drug is nicotine, followed by alcohol, heroin, cocaine, caffeine, and lastly cannabis.[13]

Trial 3 (1994) described possible drug dependence patterns as follows: most prevalent was cigarettes, next was heroin. Cocaine and alcohol followed, and cannabis ranked among the lowest.[14]

Compared to pharmaceuticals, some of which have a more significant addiction potential, cannabis carries a considerably reduced risk of adverse side effects (including death). A US Food and Drug Administration report compared marijuana to 17 common FDA-approved pharmaceutical drugs used to treat similar symptoms and conditions. These findings make a compelling argument for medical cannabis. Between 1997 and 2005, no deaths were attributed to the exclusive use of cannabis, while the FDA recorded 10,008 deaths due to the 17 FDA-approved pharmaceutical drugs in the study.[15]

Lastly, the gateway theory suggests that adolescents who experiment with cannabis are more likely to subsequently try, and become addicted to, other illicit drugs. The gateway theory has never attempted to address therapeutic uses of legally obtained medicine, but the suggestion that even short-term cannabis use could lead to addiction to other drugs still lingers in many people's minds. The facts tell a different story. A recent study of more than 4,000 cannabis smokers concluded that cannabis use leads to a decrease in alcohol, tobacco, and hard drug use and thus is an actual exit drug.[16] Medicinal cannabis could be more appropriately be named "Gateway to Health."

If you are concerned about developing a dependence on cannabis, you may reduce this potential risk by using raw cannabis-based medicines or those containing only chemotype III flower material, both of which have little or no psychoactive effect.

Lab-Tested Cannabis

To many people the concept of testing cannabis is relatively new. You may find yourself asking such questions as What is testing for? How is it relevant to me? How do I as a consumer obtain and interpret test results?

Before medical cannabis legislation, much of flower testing (if it was done at all) was for bragging rights: Whose flowers contained the most THC? However, now that cannabis has practically (re)entered the art and science of modern medicine, knowing the quantity and quality of an herbal source has become imperative for many patients and their health care providers.

Testing is contracted with a laboratory by the grower, producer, distributor, or seller to reveal what ingredients are in a cannabis flower or cannabis-containing product. States with legal access to medical or recreational use require that purveyors of cannabis (just like any food, medicine, or supplement supplier) adhere to strict quality-control standards that determine and clearly list ingredients (e.g., cannabinoids, terpenes) and potency. A passing or failing grade is given regarding the potential presence of pollution (e.g., pesticides, solvents, heavy metals), microbiological pathogens (e.g., mold, salmonella, E. coli), or filth/foreign materials (e.g., mammalian excreta). Additionally, edibles ought to be tested for homogeneity to ensure that each portion contains the same amount.

The intention behind testing is not only to avoid contaminants but to obtain (and calculate) precise amounts of individual plant constituents such as THC, CBD, or beta-caryophyllene. Similar to the use of poly-pharmaceutical drugs, specific amounts of individual ingredients, ratios to each other, as well as product purity all determine the effects, therapeutic abilities, and mitigation of possible adverse effects.

Fact is, testing becomes essential once you learn how to match your specific preferences and individual needs with the right type of cannabis, the proper dose, optimal form, and best possible combination of other ingredients.

If ingredient information is not already printed on the label, just ask for it. Most reputable dispensaries will have test results to make available for any interested or concerned consumer.

How Many Cannabinoids Are There, and Are They All Relevant?

In the latest review from 2016 (Hanuš et al.), there are approximately 150 known cannabinoids, which are generally divided into 11 groups.[17] Those listed in bold in graphic 1 are considered major cannabinoids—that is, they may occur in significant concentration to induce physiological effects in humans. Of these, THC and CBD stand out in sheer amounts (depending on cannabis types) and produce a great variety of therapeutically relevant actions of great importance for a large number of different patient populations, especially those considered chronic or degenerative. The next chapter takes a closer look at these two important cannabinoids.

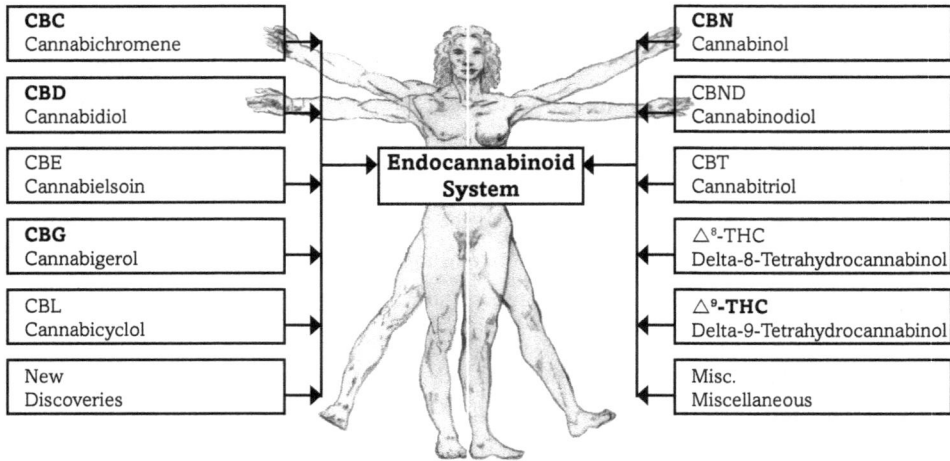

CBC Cannabichromene		**CBN** Cannabinol
CBD Cannabidiol		CBND Cannabinodiol
CBE Cannabielsoin	Endocannabinoid System	CBT Cannabitriol
CBG Cannabigerol		△⁸-THC Delta-8-Tetrahydrocannabinol
CBL Cannabicyclol		**△⁹-THC** Delta-9-Tetrahydrocannabinol
New Discoveries		Misc. Miscellaneous

Chapter 2, Graphic 1: Eleven Cannabinoid Groupings (bold/solid line arrow indicating major cannabinoid, each of which interacts with the endocannabinoid system)

Key Points, Chapter 2: Cannabis as a Healing Plant: Confusions and Insights

- Δ1-THC, Δ3-THC, Δ6a-THC, and Δ9-THC are referring to the same compound, THC.
- The greater the number of studies published relevant to a particular condition (especially if they are weighted high) indicates the degree of confidence that engaging the endocannabinoid system may be of therapeutic value.
- **The Many Names of Cannabis**

    Genus: There is only one genus of cannabis.

    Species: This genus is divided into three species: sativa, indica, and ruderalis.

    Genotype: The plant's genetic composition, or DNA blueprint of life.

    Phenotype: E.g., siblings sharing some of the same visible traits but not others.

    Hybrids: E.g., offspring from a marriage between an indica and a sativa.

    Strain: An attempt to further differentiate, or to create an additional value ranking, within one of the three prime species of cannabis.

    Marijuana: A slang term borrowed from Mexican Spanish and commonly used to denigrate the plant and its users by the narrative on the war on drugs.

    Sinsemilla: A growing technique whereby male pollen is kept away from female plants.

    Landrace: A landrace's characteristic genome or "fingerprint" of DNA emerged over hundreds or thousands of years, forged in the particular soil, weather, and light of a specific geographic region.

    Hemp: An industrial crop usually with less than 0.3% THC.

- **The Sativa versus Indica Distinctions Are Very Imprecise:** Sativa and indica plants look different in height and leaf, but it is virtually impossible to know precisely what plant constituents are actually present, and therefore what effects they may engender, by sativa versus-indica alone.
- **Biphasic Effects:** Cannabis flower, cannabis-containing products, and isolated cannabinoids can have opposite effects.
- **Fatal Dose of Cannabis:** An animal study (rats, dogs, and monkeys) suggests 200 g/kg of body weight. A 2004 review conducted at UC Washington School of Medicine concluded that "628kg of cannabis would have to be smoked in 15 min. to induce a lethal effect."
- **Adverse Effects:** Are usually dose related and most commonly include unhappiness, euphoria, increased appetite, reduced attention span, red sclera, and dry mouth. Other less common adverse effects include decreased cognitive and motor skill, ataxia, aphasia, unusual perceptions of all senses including hallucinations, anxiety, slight increase in heart rate, subtle shifts in blood pressure depending on the position of the body, and panic upon first-ever use. Some observational studies suggest that in people vulnerable to experiencing a psychotic break (e.g., people with a family history), taking a substance that can produce cognitive changes may increase the risk of schizophrenia.
- **Number of Cannabinoids:** To date, over 150 individual and chemically discernable cannabinoids have been detected and can be divided into 11 groups or families. The major ones include CBC, CBD, CBG, CBN, and THC, of which THC and CBD are currently the most relevant.

Beyond these two prime plant constituents of THC and CBD, a number of individual cannabinoids present new and exciting therapeutic possibilities such as tetrahydrocannabivarin (THCV) in the treatment and prevention of obesity, metabolic syndrome, and diabetes. Also noteworthy is cannabigerol (CBG), which has been found to produce very specific actions that include anti-inflammatory, anti-microbial, cancer-prohibitive, and muscle relaxation effects that are relevant to a number of patient populations such as those suffering from colitis, methicillin-resistant *Staphylococcus aureus* (MRSA), colorectal cancer, or patients with involuntary bladder muscle contractions, respectively. On the other hand, a number of news outlets and writers from various blogospheres have reported patients finding relief by using cannabinol (CBN) as a treatment for insomnia, a notion that is not actually supported by the currently available literature. Almost weekly, new information reveals actionable and practical suggestions emerging from ongoing research and patient-informed trends that broaden the horizon of what is possible. This cascade of scientific information about cannabis (finally!) is allowing us to make more informed decisions as we adjust and fine-tune those cannabis-based plant constituents that provide us with the precise effects we are looking for on our path to optimal health and well-being.

Chapter 3

Understanding Cannabis Pathways to Body, Mind, and Emotion

In this chapter you will learn about the following:

- Your endocannabinoid system
  - ◆ Endocannabinoid tone
  - ◆ Endocannabinoid deficiencies
- THC and CBD
  - ◆ What they are
  - ◆ What they do
  - ◆ How they do it
  - ◆ Why it matters

Humans and cannabis have been inextricably linked ever since our first primordial cannabinoid receptor evolved about 600 million years ago.[1] Since then, our very DNA has continued to make receptors that perfectly bind with plant constituents unique to cannabis.[2] In this chapter we take a closer look at the pathways that cannabis components use to connect with receptor sites embedded in a variety of cell membranes spread throughout the human body. These pathways can lead to treatments for a great number of patients suffering from chronic conditions and stubborn symptoms.

For the past several centuries, Western medicine has divided the body's systems into 11 types: the skeletal system, the muscular system, the nervous system (central, peripheral, autonomic), the digestive system (mouth, esophagus, stomach, intestines, anus), the immune/lymphatic system (blood, blood-forming organs, lymph nodes, immune cells), the respiratory system (upper airways, trachea, lungs), the exocrine/integumentary system (skin, sweat, nails, exocrine glands, hair), the cardiovascular system (heart, veins, arteries), the urinary system (kidneys, bladder), the endocrine system, and the reproductive system.

Research into each of these organ systems and their interactions with the endocannabinoid system (ECS) gives us deeper understanding and opens doors to treating many seemingly intransigent conditions. However, for far too many chronic conditions and stubborn symptoms, progress has been limited, often with no cure in sight. Too many times the best one could hope for was managing symptoms with pharmaceuticals.

This is where cannabis comes in.

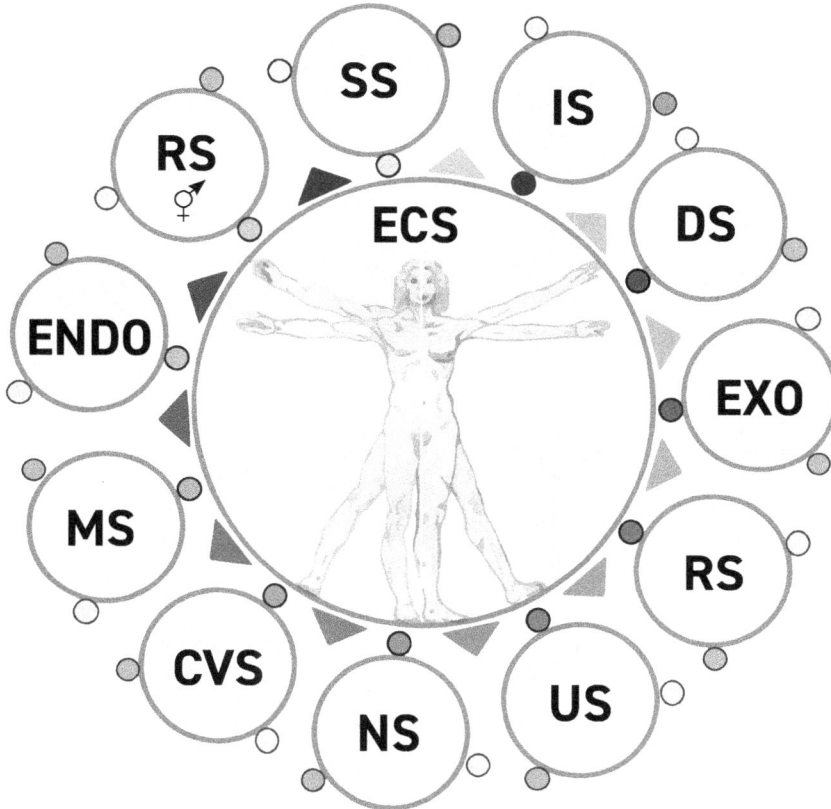

Chapter 3, Graphic 1: Eleven Organ Systems and ECS-Based Modulation

Your Endocannabinoid System

In 1990 a new physiological system was discovered—not a new organ system, but a regulatory system that either in part or significantly modulates all the classical organ systems combined. It was named the endocannabinoid system: *endo,* from the Greek word meaning "within," and *cannabinoid,* from the word *cannabis.*

Researchers looking into the natural properties of cannabis-based compounds discovered that the human body makes its own versions. That is, our bodies produce compounds very much like THC that generate similar effects in body, mind, and emotion. Compounds such as anandamide, for instance, bind to the same receptor sites as certain cannabis constituents. Anandamide is named after the Sanskrit word for bliss (think runner's high). Like other cannabinoids, anandamide can be thought of as a "key" molecule. It fits, or binds, into very specific cell receptors or locks that are distributed through all 11 organ systems.

Once the connection is made, the lock opens and a signal is generated. Now numerous physiological as well as mental and emotional changes take place that are relevant to treating chronic diseases. Anandamide initiates simultaneously a host of changes in the central nervous system, the immune system, and the autonomic nervous system, both parasympathetic (responsible for "downer" effects such as initiating rest and digestion) and sympathetic (which triggers the "upper" effect of fight or flight or freeze responses). Thus anandamide plays a supportive role in reducing pain, hypertension, depression, fear, stress, and anxiety—all important considerations in dealing with chronic conditions.

Which brings us to THC and CBD, two of the compounds that have generated a significant share of the excitement and hope born from these discoveries, two compounds that take center stage in a new and emerging field of medicine—the cannabinoid health sciences.

In many patients, especially those dealing with the impact of chronic illness, the endocannabinoid system becomes overwhelmed and weakened. As a result, it is rendered ineffective in achieving its prime purpose: multi-system homeostasis (equilibrium) that allows us to thrive. In the context of the cannabinoid health sciences, two emerging concepts attempt to explore and define this debilitating inability to regain the body's capacity for self-healing: "endocannabinoid tone" and "(clinical) endocannabinoid deficiency" (syndrome).

Exploring Endocannabinoid Tone

To better describe this relatively new (2011) concept of endocannabinoid tone, here is another example that has been studied much longer, namely vagal tone.

The vagus nerve descends from the brain's evolutionarily oldest, lowest part, the medulla oblongata. It hugs the carotid artery through the neck until it reaches the heart and abdomen. The vagus nerve is a key component of the parasympathetic/autonomic nervous system that continuously regulates the biological resting states of the heart, lungs, endocrine glands, digestive organs, and even the workings of some human psychology and behavior in any given moment.

Here are some facts about vagal tone that may help illuminate the use and meaning of endocannabinoid tone:

- Vagal tone is subjective, and it is continuous.
- It changes in direct reaction to internal and external environmental signals.
- Vagal tone aims to find balance or homeostasis. It does so by inducing rest, digestion, and relaxation in direct contrast to fight, flight, or freeze reactions.
- Vagal tone, while omnipresent, cannot be measured directly. However, its effects are clearly noticeable. For instance, a higher vagal tone slows the heart.
- Similar to the effects of anandamide,[3] higher vagal tone has been associated with psychological components such as an increased level of empathy.[4]

Endocannabinoid tone refers to the functional reach of the ECS system, which includes psychological components, while endocannabinoids are chemicals that can travel to specific locations in the human body; molecular-model research predicts that endocannabinoids are also able to interact and produce biological changes without having to travel to another receptor site. In other words, (endo)cannabinoids are able to produce change in their environment by their presence alone. This steady impact (also termed lateral diffusion) resulting from their mere presence can serve as a good working definition of endocannabinoid tone.[5] For instance, cells release anandamide under different circumstances and by different guiding mechanisms, one of which is endocannabinoid tone, as discovered by the University of Maryland.[6]

Physiologically, endocannabinoid tone is regulated in two ways: by the mechanisms of endocannabinoid production; and by their inactivation due to transport and breakdown.[7] Since THC and CBD are cannabinoids that affect the mechanisms by which endocannabinoid tone is regulated, it is likely that both may also influence, change, or shift endocannabinoid tone.

Why does it matter? A low endocannabinoid tone may be associated with a lessening of endocannabinoid-induced neuroprotective abilities—immune-modulating effects, pain reduction, and protection from oxidative stress, for example—and therefore contributes to the biological pathologies of many a chronic condition.

It matters because we may be able to take steps toward a more balanced and therapeutically supportive endocannabinoid tone. We can learn activities that support and balance our tone. And we can work with plant-based cannabinoids toward symptom reduction and improvements in mood. For instance, research has clearly shown that the body's own cannabinoid anandamide plays a role in the formation of emotion,[8] learning,[9] and resulting behavior—a fact made more fascinating by the observation that the same behavior, learning activity, and formation of emotion in turn produce more anandamide.

Anandamide release increases with the activity of high-intensity running, producing bliss or the "runner's high."[10] The same is true for playfulness,[11] meaningful social interactions,[12] learning a new skill, and certain meditation practices, all of which may increase anandamide levels naturally.

And, of course, as most cannabis-using patients will recognize, all of these tend to occur naturally with the properly dosed and responsible use of cannabis-based products.

Endocannabinoid Deficiency: When You Are Out of Balance

Endocannabinoid deficiency (or endocannabinoid deficiency syndrome) is an actual lack or insufficient production of the body's own endocannabinoids. These endocannabinoids mitigate stress responses and anxiety, for example, by reducing the impact of the excitatory neurotransmitter glutamate.[13]

A key function of the ECS is to achieve homeostasis, or an optimal balance, in response to ever-changing internal and external environments. But in some of us, chronic stressors have taxed the endocannabinoid system to the point where it is unable to produce sufficient amounts of endocannabinoids to maintain optimal homeostasis on its own. Being out of balance, we struggle to relax, be at ease, fall asleep, and maintain sleep.

The good news is that endocannabinoid deficiency can be relatively easily countered in two ways. One solution is the use of plant-based cannabinoids, and another remedy is using mindfulness techniques to reduce chronic stressors and thus support production of the body's own versions of cannabinoids.

When supplementing your ECS with plant-based cannabinoids, keep in mind that in orthodox medicine, pharmaceutical dosages are calculated based on weight; the prescribing physician or pharmacy will do this for you. In the cannabinoid health sciences, precise dosages require your own participation (trial and error, and making notes of the effects of different dosages and types). With a little awareness and participation, your results will be well worth the effort.

Tetrahydrocannabinol

Tetrahydrocannabinol (THC) is one of hundreds of cannabis constituents. More specifically, it is a cannabinoid, a term coined from the word *cannabis* and the Greek suffix "oid," indicating likeness. Its claim to fame: It can occur in significant amounts to produce a number of biological effects, and it is the primary psychoactive compound of the plant. THC is primarily (but not exclusively) responsible for many of the positive and negative attributes associated with the herb. And, of course, THC is the most studied of all the plant's individual parts. THC is exclusive to cannabis, but it is also produced synthetically.

THC is a multi-target molecule. Researchers have described more than 50 different pharmacological targets and effects of Δ9-THC.[14] A select number of double-blind, placebo-controlled, and other human trial designs highlight a few of the studies proven to benefit patients.

Chronic pain: The first double-blind placebo-controlled trial using oral THC was conducted in 1975 on 10 cancer patients with associated chronic pain. Results demonstrated effective dose levels of 15 and 20 mg (**dosages used in trial:** 5, 10, 15, and 20 mg oral THC).[15] The same year another trial conducted on 36 cancer patients with chronic pain compared the analgesic effects of THC and codeine, suggesting comparable analgesic properties (**dosages used in trial:** 10 and 20 mg THC).[16] A trial at Harvard Medical School in 2008 on 30 patients using opioids for chronic pain discovered that Marinol, a synthetic version of THC, increases the analgesic effects of opioids[17] and thus may reduce overall opioid use to produce similar analgesic effects; this reduces the potential for adverse effects, addiction, and overdose (**dosages used in trial:** 10 mg and 20 mg of Dronabinol over the course of three eight-hour visits).

In another double-blind, placebo-controlled trial, a patient with chronic pain due to chronic familial Mediterranean fever with significant gastrointestinal symptoms showed a highly significant reduction in additional analgesic requirements (**dosages used in trial:** 10 mg THC × 5 daily).[18] In 2010, 177 patients with consistent cancer pain, despite chronic opioid dosing, used an oromucosal spray containing THC and CBD in a two-week, multicenter, double-blind, randomized, placebo-controlled, parallel-group trial. Results demonstrated that the extract is efficacious for relief of pain in patients with advanced cancer pain not fully relieved by strong opioids (**dosages used in trial:** Each spray delivered a dose of 2.7 mg THC and 2.5 mg CBD and the maximum permitted dose was 8 times in any 3-hour period and 48 times in any 24-hour period.[19] By 2017 a review of the scientific literature published since 1975 identified five clinical studies that examined the effect of THC, THC with CBD, or CBD alone on controlling cancer pain. Results showed actual evidence that medical cannabis can provide effective analgesia in cases of chronic or neuropathic pain in advanced cancer patients (**dosages used in trial:** Reported THC doses ranged between 2.7 and 43.2mg/day).[20]

Chemotherapy-induced nausea and vomiting (CINV): Cannabinoids were more effective than standard anti-emetics used to treat CINV, according to a review of 30 randomized comparisons of Dronabinol or Nabilone (synthetic THC) with a placebo conducted on a total of 1,366 patients.[21]

Cancer-related loss of quality of life: This study enrolled 46 patients with advanced cancer, poor appetite, and altered chemo-perception (taste and smell) in a double-blind, placebo-controlled trial. Each was given 2.5 mg of synthetic THC (Marinol) or a placebo twice daily for 18 days. Twenty-one patients completed the trial, and results showed improved taste, smell, appetite, and caloric intake, along with increased relaxation and better overall quality of life (dosages used in trial: 2.5 mg THC twice daily).[22]

HIV-related anorexia: Dronabinol improved appetite, caloric intake, and mood in seven HIV-positive patients (**dosages used in trial:** 10 mg synthetic THC [Dronabinol] four times a day, daily total of 40 mg).[23] A double-blind placebo-controlled trial conducted on 10 HIV-positive patients similarly showed that treatment with both Dronabinol and smoked cannabis produced a dose-dependent increase in appetite, caloric intake, and weight gain (Dronabinol (5 and 10 mg) or smoked cannabis [2.0% and 3.9% THC] four times daily for four days).[24]

Tourette's Syndrome: In this randomized, double-blind, placebo-controlled trial, 24 patients with TS were treated for six weeks with daily doses of THC. Results showed that THC was effective and safe in the treatment of Tourette's (**dosages used in trial:** Up to 10 mg/day of THC).[25]

In its raw and natural state (i.e. fresh cannabis flowers), THC exists in its acid form (THCA), which is not, or only very minutely, psychoactive. When the cannabis flower is dried, heated, and exposed to light, THCA quickly converts into THC via decarboxylation, and it becomes activated or psychoactive. (See the appendix for more details on decarboxylation.)

The reader is reminded that cannabis types are most accurately distinguished by their chemotypes—that is, the ratio between the key cannabinoids THC and CBD. Type I is sometimes referred to as the "drug type" because it contains more THC than CBD. Type II contains relatively equal amounts of THC and CBD; type III, the "hemp" or "fiber" type, has small amounts (or none) of THC and more CBD.

Plant-based THC content varies by cannabis chemotype and whether it is fresh or dried (and if dried, its age), and also whether it is grown indoors or outdoors (indoor cultivation allows growers to optimize the grow environment environment—light conditions, heat, humidity—to increase cannabinoid production).

THC Ingestion vs. Inhalation: A Difference That Chemically Matters

Experienced cannabis consumers tend to have a preference and strong opinions about cannabis consumption methods. Similarly, for some newcomers, concepts such as smoking cannabis as medicine may not seem very medical and even counterintuitive—after all, smoking is bad for you, right? And while inhaling burned carbon in general is an unhealthy practice, inhaling medicine is a well-established and safe practice. Take, for example, inhalers that deliver glucocorticosteroids or bronchodilators for the treatment of respiratory illnesses such as asthma, inhalers that deliver insulin as a diabetic treatment, or inhalers that harness the benefit of bypassing the digestive tract to better deliver vitamins and supplements. One study, with practical implications, conducted by researchers at Johns Hopkins University,[26] discovered that while both smoked and vaporized cannabis containing THC produced dose-dependent effects on body, mind, and emotion, vaporized cannabinoids produced these effects in a more potent fashion. Thus it would seem that, especially for beginners, vaporized cannabis produces stronger results than smoked cannabis. Thus less medicine is needed to achieve efficacy. Therefore, it is recommended that people begin slowly with vapors to avoid adverse effects. And while inhalers containing cannabinoids are beginning to appear on the global market, this section will focus on the delivery methods of ingesting and inhaling (smoking and vaporizing).

To be clear, the absorption of THC in the human body depends greatly on the form of consumption, whether you inhale on ingest. As THC enters different tissues and organs via ingestion or inhalation, THC is broken down or metabolized. However, breakdown occurs at significantly different rates depending on whether you inhaled or ingested.[27,28] Two metabolites (breakdown products) stand out in sheer volume: 11-hydroxy-THC (aka 11-OH-THC) and THC-COOH. Of these two, 11-OH-THC is significant because it remains psychoactive and crosses the blood-brain barrier faster than THC, amplifying psychoactive effects.

Inhaled THC spreads rapidly (within minutes), and most peak effects last up to a couple of hours. In contrast, digested THC takes much longer to cause expected effects (up to two or even three hours), but when effects begin, they tend to be stronger, more intense, and longer lasting. Here is why.

Before THC, the primary psychoactive constituent of cannabis, can produce central therapeutic effects, it must enter the brain. To do so, it has to pass the blood-brain barrier (BBB), which separates most parts of the brain and spinal cord from circulating blood. The BBB restricts access to essential molecules such as glucose (the brain's primary energy source), amino acids, oxygen, and certain fat-soluble compounds.

THC is a fat-soluble molecule, so it is quite capable of passing the blood-brain barrier. While inhaled THC reaches peak concentration in the blood relatively quickly, only a portion of that crosses the BBB to initiate effects (e.g., increase appetite, relaxation).

In contrast, digested THC is very slowly absorbed into the bloodstream (an empty stomach equals faster absorption). However, when THC passes the liver, a portion of it is oxidized into an 11-hydroxy-THC that penetrates the BBB much faster than inhaled THC.[29] So, while it takes longer for ingested THC to enter and peak in blood and brain, once converted, both THC and 11-hydroxy-THC produce a synergy of speedy absorption, which leads to sudden and intense changes in body and mind (you may remember Maureen Dowd's experience from the introduction).

To summarize, inhaled cannabis bypasses the liver and thus does not create 11-hydroxy-THC. Ingested cannabis gets metabolized by the liver and produces 11-hydroxy-THC, speeding and increasing the effect of THC past the BBB once the process begins.

Here is the practical takeaway of understanding the different effects produced by inhaled versus ingested THC. Inhaled THC-containing remedies have effects that come on quickly. Effects are shorter lasting and ideal if, for example, you need help in taking a nap but must be ready to operate heavy machinery shortly after. Inhaling allows you to stop when the desired effect is achieved, even though the effect may not last as long as an ingestible form.

Digested THC-containing cannabis products are more difficult to titrate, that is, to measure the precise dose responsible for the desired effect. However, once you know your precision dose in an edible form, you can achieve effects that have a longer duration.

Taking the variables of absorption into account (such as typical loss rates during inhalation and digestion, as well as estimation of impact of 11-OH-THC), research suggests a ratio of about 1:6.[30] That is, for every 1 mg of ingested THC, about 5.7 mg of inhaled cannabis is necessary to achieve a similar effect. Keep in mind that this ratio is only an estimate, to be considered with appropriate caution. It was derived mathematically from averages from different study results, each with ranges that had large margins.

> **Remember:** This conversion ratio is a very rough guideline, and each person may react differently after trying the conversion ratio of 1:6 (ingested vs. inhaled).

Adverse Effects of THC

Can depend greatly on variables such as actual amount of cannabis constituents, form (ingestion vs. inhalation), state of your constitution (including endocannabinoid tone/deficiency), the presence of illness (and degree or stage of condition) or type of symptoms (acute vs. chronic), or the type and number of medications you are taking (the more drugs you take, the higher the risk of adverse effects).

The most common adverse effects of THC may include dry mouth, drowsiness, and mood elevation (euphoria), which are considered by a majority of patients "non-significant." Additional side effects may include dry red sclera (the white of the eye), dose-dependent anxiety (paranoia, panic attack), increased appetite ("the munchies"), sleepiness, slurred speech, sedation "couch-lock," or feeling and an unpleasant "high" (e.g., dysphoria, altered perception, or cognitive function, states of extraordinary consciousness), tachycardia (increased heart rate), and reduced blood pressure (considered a compensatory mechanism). A number of different populations may be more vulnerable to developing adverse effects, such as people with cardiovascular disease, pregnant women, patients with a history of mental illness (e.g., schizophrenia), and people driving or operating heavy machinery under the recent influence of cannabis.

General Dosage Samples for THC

To provide some context, we use the double-blind, placebo-controlled trials listed above; their range of therapeutically measurable dosages of THC (from lowest to the highest) used was oral administration of 2.5, 5, 10, 15, 20, and 43.5 mg THC. Smoked cannabis contained either 2% or ~4% THC.

Both therapeutic effects and adverse effects appeared strongly correlated with THC dosages. In general, lower dosages provided less analgesia and higher dosages provided more. Adverse effects occurred more frequently at higher dosages and not at all or to a lesser degree with lower dosages.

THC Dosage Considerations (based on dosages used in the currently available human trials)

THC micro dose: ~0.1 mg to 0.4 mg (0.001 mg/kg to 0.005 mg/kg)

THC low dose: ~0.5 mg to 5 mg (0.006 mg/kg to 0.06 mg/kg)

THC medium dose: ~6 mg to 20 mg (0.08 mg/kg to 0.27 mg/kg)

THC higher dose: ~21 mg to 50+ mg (0.28 mg/kg to 0.67mg/kg)

Mg/kg dosages calculated based on assuming a 75 kg person taking a low, medium, or high dose with the numbers as indicated (mg ÷ kg = mg/kg).

Four basic ranges—or levels of percentage by weight—of THC are commonly found in today's cultivated cannabis. Level 1 contains up to 7% THC by weight, level 2 is 8–14% by weight, level 3 is 15–21% by weight, and level 4 between 22 and 28+%. However, there are a few strains (as of 2016) that surpass 30% (Godfather OG, ~34%; Super Glue, ~32%; and Strawberry Banana, ~32%).[31] Later on in this book you will learn how to easily calculate actual THC mg content from the plant's percentage amount.

THC Interaction with Pharmaceutical Drugs

THC is a molecule that binds with receptor sites throughout all human organ systems, so it has the potential to partially or significantly modulate the action of a great number of foods, drinks, and drugs (i.e. enhancing effects, decreasing effects, or creating a synergy of effects). However, much remains to be learned about the specific ways THC might interact with other drugs. As always, discuss the interaction potential with a licensed health care provider well versed in the cannabinoid health sciences. Here are some general concerns and considerations that may help you to take a more informed or discerning view.

THC can enhance the effects of drugs that cause sedation and depress the central nervous system, such as benzodiazepines, barbiturates, and alcohol. THC is metabolized by and an inhibitor of a number of enzymatic liver pathways referred to as cytochrome P450.[32] There are more than 50 enzymes belonging to this enzyme family, a number of which are responsible for the breakdown of common drugs such as antidepressants (e.g., amitriptyline, doxepine, fluvoxamine), antipsychotics (haloperidol, clozapine, stelazine), or beta-blockers (propranolol, theophylline, warfarin).[33] Thus patients taking these classes of medication may find that THC increases the concentration and effects of these drugs as well as the duration of their effects.

Cannabidiol

CBD is the other major cannabinoid that occurs naturally in cannabis in amounts large enough to produce various and significant changes in the human body, mind, and emotions. CBD also exists in synthetic forms.

CBD is, like THC, a multi-target molecule that binds with numerous cell receptors throughout the body; thus it directly and indirectly initiates complex biological changes. One study highlighted more than 30 pharmacological actions for CBD.[34] Here are a few of them.

A select pre-clinical trial summary suggests that CBD initiates activation of serotonin (5HT1A) receptors,[35] lowers heart rate,[36] produces mood improvements,[37] and aids sexual arousal.[38] CBD is also considered a potentially potent anticonvulsant for various disorders involving spasms, fits, convulsions, seizures, and forms of epilepsy.[39–44]

Two examples: One, a plant-based extract named Epidiolex that contains only CBD, is now FDA-approved for numerous chronic neurological disorders such as treatment-resistant epilepsy in infants and children, Dravet syndrome, and Lennox-Gastaut syndrome. Two, Sativex, a cannabis-based extract containing THC and CBD in a roughly 1:1 ratio, is approved in about 30 countries for the treatment of pain and spasms in patients with multiple sclerosis.

CBD's New Promises in Healing

Additionally, CBD has caused excitement in the research community as a novel and promising agent in the co-treatment of pain in conjunction with opioids. Animal studies have shown that CBD can boost opioid-based analgesic effects, allowing for lower doses that are effective, while reducing the risk of addiction and overdose. For instance, German researchers discovered in rodent experiments that both CBD and to a lesser degree THC (six times less than CBD) bind non-competitively with mu (μ) and delta (δ) opioid receptors,[45] and thus they may potentially enhance the effects of opioids. If confirmed in humans, this synergy could explain why many patients using opioids for the treatment of pain tend to use significantly less when also medicating with cannabis.

CBD's further potential properties include neuroprotection in cases of Alzheimer's[46] and Parkinson's diseases,[47,48] potentially protective against MRSA,[49] and inhibitive against certain cancers (e.g., lung[50] and breast[51]). Studies also show the effects of CBD to be neuroprotective and antioxidant—which makes it of value to numerous patient populations, especially those suffering from chronic diseases with underlying pathologies of oxidative stress, neurodegeneration, aging, ischemia, inflammatory states, or immune disorders, for instance.[52–57]

CBD has been abundantly studied in double-blind trials containing both THC and CBD (in a 1:1 ratio), especially in the context of chronic degenerative neurological disorders such as MS.[58–61] In contrast, double-blind CBD-only trials have been more limited. Following is a summary list of the existing double-blind trials that tend to confirm many of the emerging treatment trends highlighted in the pre-clinical domain: chronic pain,[62,63] cardio-vascular support,[64] mood improvements.[65–8] CBD also produced a reduction in anxiety and improved memory (recall), and contributed to a calming effect in the presence of stress, along with a decrease in intravenous THC-induced psychotic symptoms.[69–71] CBD has been found to have anti-addictive (nicotine)[72] properties, relevant to the opioid epidemic, and has been shown to support patients dealing with insulin resistance and obesity.[73]

Adverse Effects of CBD

A 2011 review of 132 scientific studies describes the safety profile of CBD as follows: "Catalepsy is not induced and physiological parameters are not altered (heart rate, blood pressure, and body temperature). Moreover, psychological and psychomotor functions are not adversely affected. The same holds true for gastrointestinal transit, food intake, and absence of toxicity. Chronic use and high doses of up to 1500 mg per day have been repeatedly shown to be well tolerated by humans."[74]

An updated review (2017) continues to confirm the safety parameters highlighted in the original.[75]

Determining the precise dosage of a cannabinoid such as CBD is a subjective process that greatly depends on a number of subjective factors: pain (physical, mental, and/or emotional), diet, type and stage of your condition, presence and number of pharmaceutical drugs, accumulation of toxins, and loss of functions. These chronic stressors tax our natural stress-buffering mechanism of the endocannabinoid system to the point where it is unable to produce sufficient amounts of endocannabinoids to maintain optimal homeostasis on its own.

Being out of balance, we struggle to relax, be at ease, fall asleep, and maintain sleep. When we introduce an external cannabinoid such as CBD, we can begin to address the imbalance and strengthen our natural endocannabinoid tone or capacity to return to homeostasis in all organ systems.

General Dosage Samples for CBD

Keep in mind that in orthodox medicine, pharmaceutical dosages are calculated based on a patient's weight. The prescribing physician or pharmacy will do this for you. In the cannabinoid health sciences, precision dosages may require your participation.

As the trials have shown, ingestible CBD products (devoid of pesticides, toxins, or organic contaminations) have been safe to consume both for healthy volunteers and for patients dealing with chronic health challenges such as Parkinson's disease. However, your current state will greatly determine your need. To determine your optimal dose, start low and slow, and gradually increase your dosage per day to where you realize the desired symptom reductions or mood improvements.

To provide you with some context from the double-blind, placebo-controlled trials, here is the range of therapeutically measurable dosages of CBD used: 0.4 mg, 2.5 mg, 16 mg, 32 mg, 75 mg, 100 mg, 300 mg, 400 mg, 600 mg, and 800 mg. Here are some approximate threshold guidelines derived from the double-blind trials listed above:

CBD Dosage Considerations (based on dosages used in the currently available human trials)

CBD low dose: 0.4 mg to 19 mg (0.005 mg/kg to 0.25 mg/kg)

CBD medium dose: 20 mg to 99 mg (0.26 mg/kg to 1.32 mg/kg)

CBD high dose: 100 mg to 800+ mg (1.33 mg/kg to 10.7 mg/kg)

Mg/kg dosages calculated based on assuming a 75 kg person taking a low, medium, or high dose with the numbers as indicated (mg ÷ kg = mg/kg).

CBD Interaction with Pharmaceutical Drugs

Pre-clinical and clinical trials conducted on humans suggest that CBD (especially at higher dosages) may alter action on metabolic enzymes (compounds that digest a substance), and thus may alter interactions with other drugs, some of which may produce enhanced therapeutic or adverse effects. For instance, CBD interacts with a number of members of the enzyme family cytochrome P450, increasing the bioavailability of anti-epileptic drugs such as clobazam (a benzodiazepine). This makes it possible to achieve the same results at significantly lower dosages, reducing treatment costs and risks of adverse effects.[76] Groups of drugs commonly affected include anti-epileptic drugs, a number of psychiatric drugs, and drugs affecting metabolic enzymes, for example. More studies are underway to better understand these various mechanisms.

As you can see, each of these cannabinoids alone have significant potential to produce a great variety of therapeutic effects for the human body, mind, and emotions. The next section examines the pathways by which THC and CBD initiate their multi-faceted effects. These pathways become the scientific basis for turning your cannabis prescription into precision medicine.

The Understanding of the Cannabinoid Health Sciences in the Early 1990s

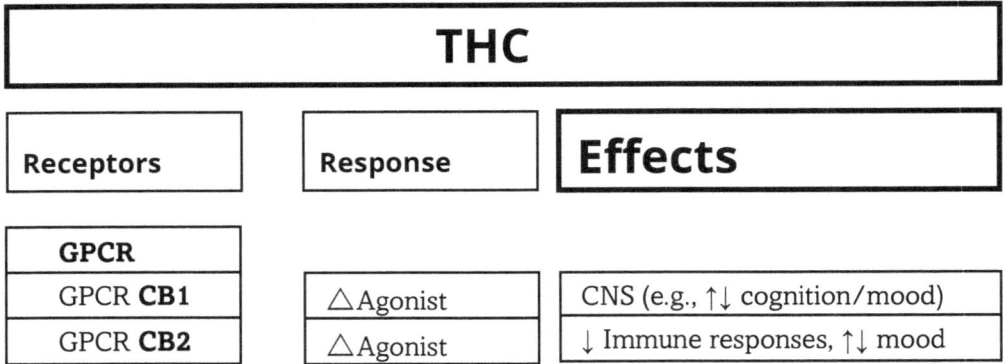

THC		

Receptors	Response	Effects

GPCR		
GPCR **CB1**	△ Agonist	CNS (e.g., ↑↓ cognition/mood)
GPCR **CB2**	△ Agonist	↓ Immune responses, ↑↓ mood

THC and CBD Light Up Pathways in Your Body

By 1990 cannabis research had recognized the existence of the classical receptor sites, cannabinoid receptors 1 and 2 (CB1 and CB2). The general narrative of that period defined CB1 receptors as affecting the central nervous system (responsible for psychotropic effects), and CB2 receptors as affecting the peripheral nervous system and immune systems of the body (non-psychotropic). Researchers had also discovered that the human body makes its own version of cannabis (called endocannabinoids), some of the mechanisms enzymes use to break it down, and what effects would ensue. However, the emerging cannabinoid health sciences were only beginning to emerge and thus were very limited in the complexity needed to make more informed and evidence-based decisions about which cannabis constituents would be optimal for which patient population. Graphic 3 in this chapter shows the relatively small scope of scientific understanding on the ECS at that time.

CBD		
Receptors	**Response**	**Effects**

GPCR		
GPCR **CB1**	△ Agonist	CBD ↓ effect of THC
GPCR **CB2**	△ Agonist	↓↑ Immune, ↑ mood
Enzymes	↓ FAAH ↑ AEA	↑ Mood & ↓ Psychosis

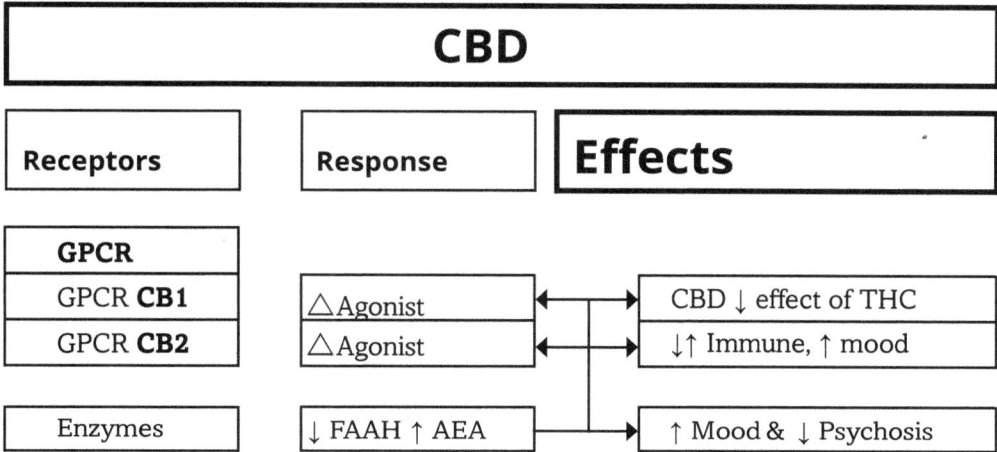

Chapter 3, Graphic 2: The Basic Understanding of the ECS (1990)

The top row shows the prime cannabinoids, THC on the left and CBD on the right side of the two-page spread. Directly beneath are the left columns that contain receptor sites that bind with THC and CBD. The center colums on each page represent the types of responses between THC (a moderate agonist at CB1 and CB2) and CBD (a very weak agonist at CB1 and a weak agonist at CB2) and specific receptor sites, and the columns on the right side on either page are examples of specific effects that are being generated.

Since then, the amount of new research results has expanded very rapidly. The currently available scientific evidence provides us with a much more complex understanding far beyond the rudimentary basis just a few decades ago (see chapter 3, graphic 3). This is indeed very exciting news for patients and for physicians alike. It is this new complexity that gives rise to the importance of understanding the ratios between the plant's primary cannabinoids, which is one of the main reasons why creating and sustaining precision effects are now more possible than ever before.

Summary of Effects Produced by THC and CBD

THC		

Receptors	Response	Effects
TRPs		
TRPA1	△ Agonist[2]	May ↓ pain, itch, & obesity[1]
TRPV1	N/A	N/A
TRPV2	△ Agonist[4]	May ↓ heat, pain, inflammation[3]
TRPV4	N/A	N/A
TRPM8	▽ Antagonist[6]	May ↓ pain[5]
GPCR		
Opioid γ/δ	(+)Allo. mod.	Synergistic analgesia (6×<CBD)[7]
GPCR **CB1**	△ Agonist	CNS (e.g., ↑↓cognition/mood)
GPCR **CB2**	△ Agonist	↓ Immune responses,[8] ↑↓ mood[9]
GPCR 3	N/A	N/A
GPCR 6	N/A	N/A
GPCR 18	△ Agonist	Testes, spleen,[10] may ↓ high BP[11]
GPCR 55	△ Agonist[17]	Heart funct.,[13] may ↑ risk for ca.[14-16]
Ligands	(+/-) Allo. mod.	↑ positive affect/mood
PPARs γ	△ Agonist γ[20,21]	May ↓ hypertension,[18] ↓ cancer[19]
Enzymes	N/A	N/A

N/A not applicable
↑ Increases
↓ Decreases
△ Agonist (produces action)
▽ Antagonist (weakens/blocks action)

CBD

Receptors	Response	Effects
TRPs		
TRPA1	△ Agonist[22]	May ↓ pain, itch, and inflammation
TRPV1	△ Agonist[23,24]	May ↓ pain[25]
TRPV2	△ Agonist[26]	May ↓ chronic pain[27]
TRPV4	△ Agonist[28]	May ↓ acne[29]
TRPM8	▽ Antagonist[30]	↓ Inflammation,[31] ↓ prostate ca.[32]
GPCR		
Opioid γ/δ	(+)Allo. mod.	↑ Synergistic analgesia (6×>THC)[33]
GPCR **CB1**	△ Ago.[34] ▽ Anta.*[37]	CBD ↓ effect of THC[35,36]
GPCR **CB2**	△ Ago.[38] ▽ Anta.*[41]	↓↑ Immune,[39] ↑ mood[40]
GPCR 3	◇ Inverse agonist[42]	↓ Alzheimer's,[43] neuropathy[44]
GPCR 6	◇ Inverse agonist[45]	↓ Parkinson's disease[46]
GPCR 18	N/A	N/A
GPCR 55	▽ Antagonist[47]	↓ Bone loss,[48] may ↓ certain cancers[49–51]
Ligands	(+/-) Allo. mod.	e.g., ↑ 5HT1A[52] ↑ GABA$_A$[53] ↓ Adenosine[54]
PPAR's γ	△ Agonist[55]	↓ Lung ca.,[56] ↓ inflammation & pain[57,58]
Enzymes	↓ FAAH[59] ↑ AEA	↑ Mood & ↓ Psychosis[60]

◇ Inverse agonist (opposite action)
(+) Positive allosteric modulator (amplifies action of an agonist)
(-) Negative allosteric modulator (diminishes action of an agonist)
δ-delta α-alpha, μ-mu, γ-gamma
*Selective allosteric antagonist of CB1/CB2 receptor agonists

Chapter 3, Graphic 3: The More Complex Understanding of the Endocannabinoid System (2020) Is the Scientific Basis for the Chemotype-Based Use of Cannabis

Legend: Summary Effects (Partial) Produced by THC and CBD

In the left colums on either page you will see blocks/groups of various receptor sites.

- **1. Block:** Ionic cannabinoid receptors aka transient receptor potential (TRP) channels pronounced "trips" include TRPA1, TRPV1, TRPV2, TRPV4, and TRPM8 ("A" for stands ankyrin), ("V" stands for vanilloid), and ("M" for stands melastatin).
- **2. Block:** G-protein coupled receptors (GPCR) include GPCR CB1 aand CB2, GPCR 3, GPCR 6, GPCR 18, GPCR 55, and opioid receptor sites (mu/μ and delta/δ).
- **Block 3:** Various ligands and their cannabinoid-sensitive receptor sites modulation via (+ or -) allosteric mechanisms (see list in chapter 3, graphic 6).
- **Block 4:** Nuclear receptor proteins called peroxisome proliferator-activated receptors or PPARs (pronounced "peepars").
- **Block 5:** Enzymatic pathways of endocannabinoid degradation (e.g., anandamide, 2-AG)

In the **center columns** on either side of the page you will find what type of interactions either THC or CBD produces at each of the receptor sites. Generally speaking, they can either interact as an agonist ($\triangle$), as an antagonist (∇), as an inverse agonist ($\lozenge$), or as a (+ or -) allosteric modulator.

- **Agonist** ($\triangle$): Interacts to produces a specific effect.
- **Antagonist** (∇): Interacts to block or reduce the normal effect.
- **Inverse agonist** ($\lozenge$): Interacts to create the opposite effect.
- **Allosteric modulator** (+ or -): Allosteric binding sites are indirect binding sites that can influence the effect of an agonist or antagonist. (In contrast, orthosteric binding sites are direct or primary binding sites of a receptor to a compound.)

And finally, the right-side columns on either page provide the reader with a few examples of the specific effects that are generated by either THC or CBD.

```
┌─────────────────────┐        ┌──────────────────────────────────────────┐
│                     │◄───────│ (+) Positive allosteric (indirect) modulator:│
│ Allosteric (indirect)│       │ e.g., CBD amplifies opioid receptor action │
│ binding sites        │◄──┐   └──────────────────────────────────────────┘
└─────────────────────┘   │   ┌──────────────────────────────────────────┐
                          └───│ (-) Negative allosteric (indirect) modulator:│
                              │ e.g., WIN 55212 inhibits acetylcholine      │
                              └──────────────────────────────────────────┘
```

Orthosteric (direct) binding site	△ Strong agonist (1-9nM) e.g., JWH-018 @CB1 (~9nM)
	△ Moderate agonist (10-99nM) e.g., THC @CB1 (~25nM)
Cell membrane	△ Weak agonist (100-999nM) e.g., Anandamide @CB1 (~240nM)
	△ Very Weak agonist (1000+nM) e.g., CBD @CB1 (~1,500nM)
	▽ Antagonist: weakens/blocks action e.g., CBD @GPCR55 (~445nM)
e.g., 7 membrane G-protein coupled receptor sites (CB1, CB2, GPCR 3, 6, 18, 55, opioid α, γ)	◇ Inverse agonist: opposite effect e.g., SR-141716 @CB1 (2.9nM)

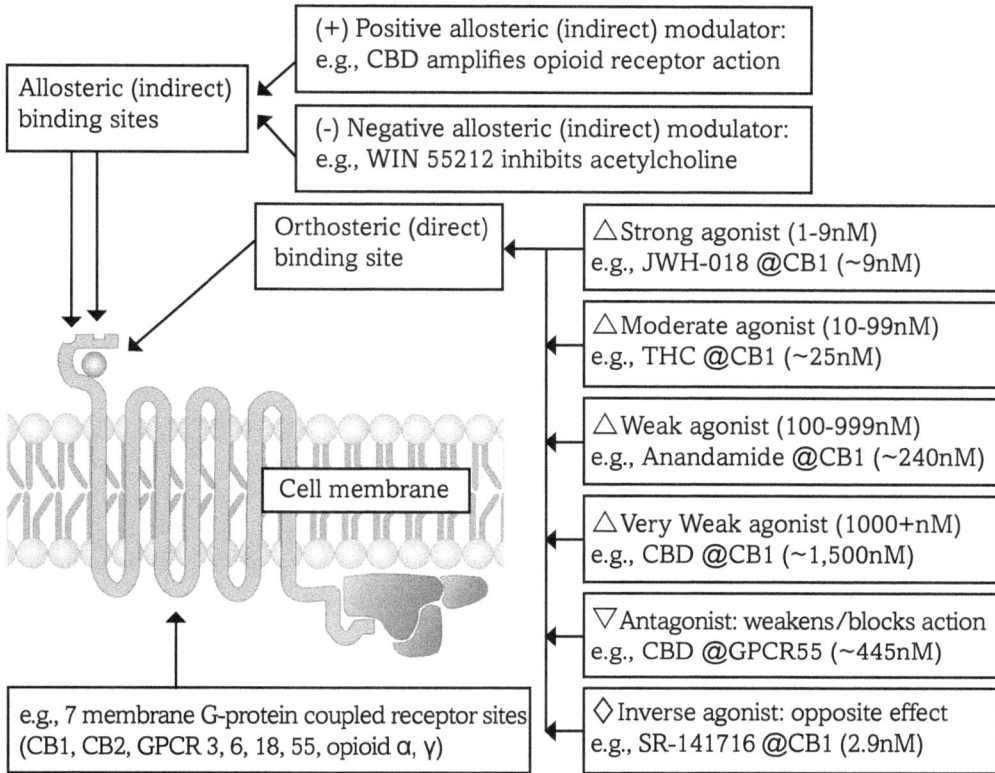

Chapter 3, Graphic 4. Types and strength of various cannabinoids at various endocannabinoid receptor sites. The strength with which an agonist such as THC binds with a specific receptor determines the type and duration of the effect it will produce (the efficacy), and it is measured in a unit of measurement called a nano Molar (nM), which stands for a factor of one billionth or 0.000,000,001 or 10-9. This graphic contains a number of examples of nM measurements. The reader is reminded that the larger the nM number, the weaker the effect, and vice versa.

As you might imagine, the far-reaching impact of working with the ECS can be key to a much more complex, complete, and deeper approach to healing. This is especially true for chronic illness and stubborn symptoms whose origins orthodox medicine does not fully understand, where no mainstream cure exists, and where managing multiple symptoms with more pharmaceuticals is the standard model of care. For instance, have you ever wondered why various cannabinoids that bind to the same receptor site such as CB1 can produce a great variety of mental and emotional effects that range somewhere from organ failure and death to producing gentle therapeutic effects in the treatment of mood disorders? The answer is that while various cannabinoids bind to the same receptor (CB1), they do so with different intensity or strengths. As a result, you will get the synthetic cannabinoid JWH-018 commonly used in "spice" that binds with CB1 at full force (**strong/full agonist**) to produce the many dangerous effects you will hear about in the news when people have succumbed to the many ill effects of this potent synthetic cannabinoid. You will get the plant-based cannabinoid

THC that binds to CB1 with a moderate force (moderate agonist) and, depending on dose and sensitivity of the person using it, can produce cognitive/emotional changes that range between a pleasant "high" and anxiety, for example, but with no risk of of organ damage or death. Next is anandamide, the body's own even milder version of THC that is a weak agonist at CB1 and will not produce changes in cognition but is still quite able to produce positive emotional changes (think runner's high). And, lastly, there is CBD, which happens to be a **very weak agonist** at CB1. It will not produce any changes in cognition (no "high"), but because it diminished the action of the enzyme (FAAH) that breaks down anandamide, it increases the bioavailability of the endocannabinoid and thus is fully capable of producing the mood-improving effect that so many patient populations have come to rely on.

With this in mind, let us take a deeper look at your body's receptor sites that are activated by cannabinoids and the effects they are capable of producing.

Your Body's ECB Receptors

Both THC and CBD bind to a number of unique receptor sites—which, once engaged, produce a unique signal that in turn creates a specific physiological and mental/emotional response. By understanding even some of these connections, we can better harness a specific therapeutic mechanism. The following pages contain a brief discussion of these five prominent blocks of receptor sites or mechanisms by which THC and CBD produce their significant effects.

The graphic may seem complex and even intimidating, but you do not have to remember all the details and their meanings,. This book is designed to help you in a step-by-step process to make more discerning decisions based on evidentiary data behind the graphic.

The idea here is to understand that THC and CBD do produce a great number of effects in and all by themselves—but also in relationship to each other. For instance, sometimes THC and CBD produce similar effects but by different mechanisms. You can use the chemotype graphics as a reference to make more discerning decisions and to create more precise effects. This will make a difference when you need to reduce inflammation but need to operate heavy machinery and thus cannot get "high." Other times it is helpful to know how to create a synergy that is an effect larger than each could have achieved by itself—helpful knowledge when trying to work in combination with opioids to produce potent yet safer pain-reducing analgesia. Yet other times one cannabinoid tames or reduces the effects of the other; you may need to produce appetite and want to make sure you are not making your anxiety worse, in which case you'll want to make sure your remedy includes the optimal ratio of THC:CBD that is right for you.

Here is a summary and a few more practical examples by which THC and CBD initiate their distinct therapeutic effects.

Ionotropic cannabinoid receptor sites (1. Block)

Ionotropic cannabinoid receptors (ICR), commonly referred to as transient receptor potential ion channels (TRP, pronounced "trip" or "trips"), are embedded in cellular membranes in tissues throughout the human body. The key thing to remember with TRPs is the name ion, indicating the involvement in receptor action by an electrical charge.

Cannabis-based constituents such as THC and CBD (and a number of cannabis-based terpenes) bind with several TRPs. THC binds with TRPA1, TRPV2, and TRPM8; in contrast, CBD binds with TRPA1, TRPV1, TRPV2, and TRPV4. Take note that CBD antagonizes (blocks or diminishes) TRPM8 and thus is in direct opposition to THC.

TRPA1 (A = ankyrin) is highly expressed in sensory nerves and believed to affect patients dealing with pain, infection, immune challenges, and neurogenic inflammation.[77]

TRPV1 (V = vanilloid) is expressed in sensory nerve cells and is widespread throughout the central nervous system (also eye, nose, skin, blood, and vascular cells). Among its functions are sensation and determining sources of heat, as well as regulating body temperature.[78] Activation and antagonism (blocking) of this type of receptor may produce responses relevant to inflammatory pain and neuropathic pain, respectively. TRPV1 antagonism may lead to hypothermia. Their role in vascular responses[79,80] has made TRPV1 receptors an attractive target for researchers seeking new treatment options for chronic hypertension and cardiovascular conditions. TRPV1 receptors may be an attractive target for reducing obesity. Research conducted on animals[81] and humans demonstrates that TRPV1 modulation significantly reduces obesity.[82] CBD has been shown to slow metastatic lung cancer proliferation by 83%.[83,84]

TRPV2 is expressed in immune cells such as macrophages.[85] These receptors are activated by physical stimuli such as stretching, heating, and osmotic swelling, and chemical modulators such as the cannabinoid CBD. They may play a role in the treatment of chronic pain, especially pain of an inflammatory nature. TRPV2 is also therapeutically involved in infectious conditions and immune-system regulation. For instance, one study determined that TRPV2 receptors are expressed in skin cells affected by a common and chronic inflammatory skin condition called rosacea, and may be a target for new treatments.[86]

TRPV4 receptors are expressed in skin cells. CBD-based activation of TRPV4 inhibits sebocyte lipogenesis and thus may present a novel approach to treating acne.[87]

TRPM8 (M = melastatin) receptors are highly expressed in a large variety of tissues including the human central and peripheral nervous systems, prostate, and liver.[88] TRPM8 is activated by the sensation of cold, cold-based pain, and cooling chemicals such as menthol. THC (also testosterone) are agonists at TRPM8. In contrast, CBD is an antagonist at TRPM8. Since it has been discovered that TRPM8 receptors exist in certain malignant tumors (including carcinomas of the prostate, pancreas, and breast), they are receptor sites considered to be novel targets for therapeutic modulation. For instance, TRPM8 receptors are required for pancreatic cancer cells to proliferate.[89] In contrast, an animal study showed that CBD inhibited the growth of prostate cancer by 34% through antagonistic targeting of TRPM8.[90] In other words, CBD was found to be partially responsible for the destruction (apoptosis) of prostate cancer cell lines and thus may present a new treatment potential for this condition.[91]

G-protein-coupled receptor sites (2. Block)

G-protein-coupled receptor (GPCR) sites are named after guanine, a nucleotide building block that makes up the sequences of our DNA (the carrier of all of our genetic potential). Many hundreds of different types of these receptors are inside the human body, and one function of GPCR is to respond to the presence of environmental signals such as light, aromatic olfactory pheromones, and molecules that elicit sweet and bitter tastes. Another big function is to respond to endogenous neurotransmitter and hormone molecules. Once a particular signal is detected, the cell receptor responds in a number of ways that support life and well-being. Malfunction of GPCR can cause serious chronic conditions and death. A number of GPCR types respond to cannabis.

CB1, CB2: These are classical cannabinoid receptors. Initially researchers believed that CB1 receptors were isolated to the brain and central nervous system and CB2 receptors primarily in cells of the periphery and the immune system. Later findings showed that both may be present at different concentrations on both sides of that generalized central versus peripheral/immune division. This is especially true under specific conditions such as inflammation. Under inflammatory conditions, CB2 expression is enhanced in the brain to produce significant anti-inflammatory effects.[92] These findings are especially relevant to patients dealing with neuro-inflammatory conditions (Alzheimer's, cerebral ischemic/reperfusion injury), neuropsychiatric issues (depression, schizophrenia), and neuro-degenerative disorders (ALS, MS).

There is evidence that GPCR 3, 6, 18 and 55 are responding to cannabinoids as well.

GPR-3: Recent data implicate the GPR-3 in the development of Alzheimer's disease. CBD may provide a new therapeutic approach to mitigate Alzheimer's sysmptoms and slow disease progression.[93]

GPR-6: Similarly, research results suggest that GPR-6 is involved in the development of Parkinson's disease. CBD may hold promise for a novel therapeutic approach to Parkinson's disease prevention or the mitigation of its effects.[94]

GPR-18 contributes significantly to the production of self-renewing cells (microglia) and prevents the build-up of pro-inflammatory microglia that underpin numerous neuro-degenerative diseases such as multiple sclerosis.[95]

GPR-55 is involved in energy metabolism (through the regulation of food intake and appetite), fuel storage in fat cells, gut motility, and insulin secretion. Given these relationships, modulating GPR-55 may represent a new target in the treatment of obesity and type-2 diabetes.[96] Additionally, activation of 55 has been implicated in the proliferation of certain potentially metastasizing cancers (e.g., ovarian, breast, colon).[97-99] THC is an agonist at 55. In contrast, CBD is an antagonist at 55. While much remains to be learned about the dynamics of both THC and CBD on GPR-55 in the context of cancer, CBD-based antagonism may prove a novel element in considering the optimal treatment options.

Opioid receptors

Like cannabinoid receptors CB1 and CB2, opioid receptors are also G-protein-coupled receptors. The primary molecules that bind to these sites are opioids from exogenous (outside the body) sources such as the pharmaceutical drug morphine and the plant-based (poppy) opium, or opioids produced by the body itself (endogenous) such as endorphins. Endogenous opioids bind with opiate receptors to reduce stress and pain perceptions; they are also involved in the formation of emotions (elation, euphoria), interpersonal relationships (intimacy, trust), and hunger.

While opium-based analgesia has been used for thousands of years, opioid receptor sites were not discovered by modern science until the early 1970s. Candace Pert, who would later write the book *Molecules of Emotion* (1997), and Solomon Snyder are the principal investigators credited with the discovery of the first opioid receptor site, the mu (μ).[100]

> While the terms are becoming interchangeable today, the word *opiate* initially referred to plant-derived medicines in this family; the term *opioids* was adopted later to distinguish artificial (synthetic) versions of the poppy plant's natural medicine.

Both CBD and to a lesser degree THC bind non-competitively with mu (μ) and delta (δ) opioid receptors,[101] and thus may potentially enhance the effects of opioids, as shown by research conducted at the University of Bonn in 2006 on rat brains. If confirmed in humans, this synergy could explain why many patients using opioids for the treatment of pain tend to use significantly less when also medicating with cannabis.

Numerous statistical reports,[102] patient-based case studies,[103] and surveys by patient advocacy groups[104] have confirmed a synergistic link between the use of cannabinoids and opioids. In other words, a combination of both may allow for opioid treatment at lower doses, with significantly fewer side effects, significantly fewer overdose deaths, less addiction, and a much lower cost. This is an extremely promising development for the US national opioid crisis.

More studies will provide further insights into how cannabinoids specifically influence analgesic effects by binding allosterically with opioid receptor sites. It would seem prudent—especially given the positive safety record of cannabis-based medication—to allow patients suffering from pain, particularly chronic pain, safe access wherever and whenever possible.

Ligands and their cannabinoid-sensitive receptor sites (3. Block)

Both THC and CBD create changes by modulating (e.g., +/- allosteric) various ligands such as neurotransmitters, hormones, signaling proteins, analgesics, and their associated affectual states or emotions. For instance, CBD suppresses the enzyme that breaks down anandamide, the body's own version of THC. As a result, the "bliss molecule" is bioavailable for longer durations and so are the effects it engenders, such as a gentle uplift in emotion, a sense of social connection, and analgesia.

The chart (chapter 3, graphic 5) illustrates common ligands and samples of their corresponding emotions that are partly or significantly modulated by the endocannabinoid system.

The Endocannabinoid System is the Biological Basis for Mind-Body Medicine[1]

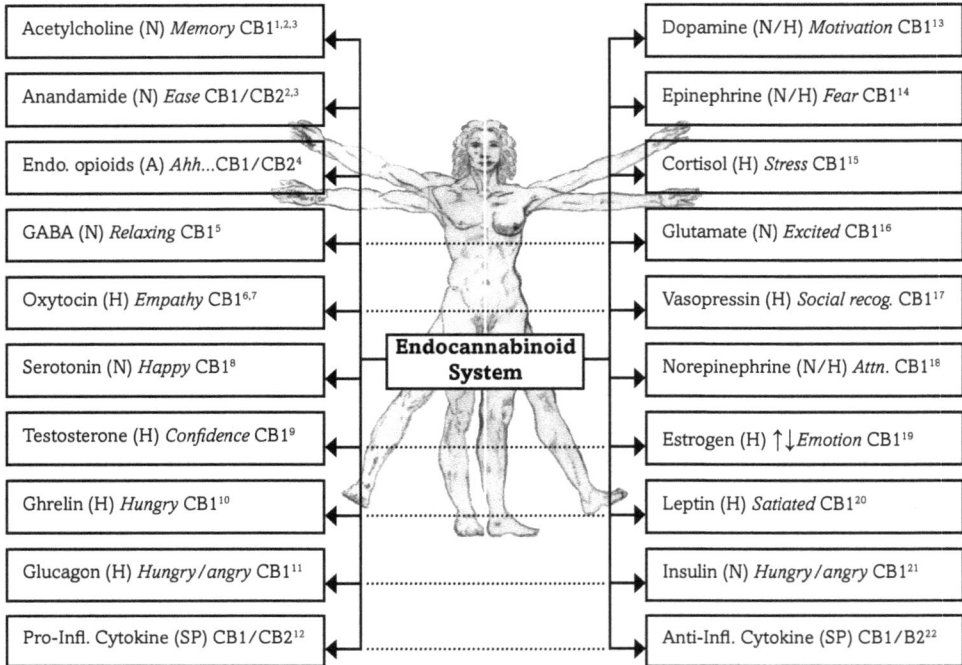

Acetylcholine (N) *Memory* CB1[1,2,3]	Dopamine (N/H) *Motivation* CB1[13]
Anandamide (N) *Ease* CB1/CB2[2,3]	Epinephrine (N/H) *Fear* CB1[14]
Endo. opioids (A) *Ahh*...CB1/CB2[4]	Cortisol (H) *Stress* CB1[15]
GABA (N) *Relaxing* CB1[5]	Glutamate (N) *Excited* CB1[16]
Oxytocin (H) *Empathy* CB1[6,7]	Vasopressin (H) *Social recog.* CB1[17]
Serotonin (N) *Happy* CB1[8]	Norepinephrine (N/H) *Attn.* CB1[18]
Testosterone (H) *Confidence* CB1[9]	Estrogen (H) ↑↓*Emotion* CB1[19]
Ghrelin (H) *Hungry* CB1[10]	Leptin (H) *Satiated* CB1[20]
Glucagon (H) *Hungry/angry* CB1[11]	Insulin (N) *Hungry/angry* CB1[21]
Pro-Infl. Cytokine (SP) CB1/CB2[12]	Anti-Infl. Cytokine (SP) CB1/B2[22]

Endocannabinoid System

Chapter 3, Graphic 5: Communication Molecules Modulated by the Endocannabinoid System (N) Neurotransmitter, (H) Hormone, (N/H) Neurotransmitter Hormone, (A) Analgesic, (SP) Signaling Protein. The commonly associated emotion(s) or mental function(s) are indicated in italics.

Peroxisome proliferator-activated receptors (PPARs) (4. Block)

You can think of the peroxisome as a cellular "organ" contained in human cells that is responsible for numerous functions, one of which is the breakdown of long chains of fatty acids and reactive oxygen species. PPAR activation via THC or CBD can have therapeutic effects for patients with diabetes,[105,106] obesity,[107] and certain cancers.[108,109] THC and CBD are agonists at PPAR-γ (gamma). CBD facilitates PPAR-alpha activity by inhibiting the enzyme FAAH.

Endocannabinoid metabolizing enzymes (5. Block)

The first of the body's own endocannabinoids were discovered in the early 1990s, but their exact mechanism for enzyme breakdown was not known until 2002, when researchers found the crystalline structure of the enzyme fatty acid amide hydrolase (FAAH).[110] FAAH is the enzyme produced and regulated by the human body that breaks down anandamide—and thus eliminates its biological effect on CB1 and CB2 (and possibly elsewhere).

Animal tests have shown that suppressing the enzyme FAAH significantly increases anandamide, thus sustaining anandamide bioavailability and signaling in the body via CB1 and CB2—a mechanism coinciding with significant analgesic effects.[111-114] Furthermore, suppression of FAAH has shown relevance in the treatment of certain stress and anxiety disorders[115] such as post-traumatic stress disorder (PTSD),[116] cognitive decline, and drug addiction.[117]

CBD inhibits FAAH and thus may have analgesic, anxiolytic,[118] stress-reducing,[119] and otherwise supportive abilities relevant to very specific patient populations such as those suffering from schizophrenia.[120]

Key Points, Chapter 3: Understanding Cannabis Pathways to Body, Mind, and Emotion

- **The endocannabinoid system** is a regulatory system that partially or significantly modulates all classical organ systems.
- **Targeted ECS** activation can become a novel way to therapeutically support many conditions, illnesses, and symptoms associated with each target.
- **Endocannabinoid tone** refers to the functional reach of the ECS system. (Endo)cannabinoids are able to produce change by their presence alone.
- **Endocannabinoid deficiency** (or clinical endocannabinoid deficiency syndrome) can be thought of as the lack or insufficient production of the body's own endocannabinoids.
- **Why do both concepts matter?** A low endocannabinoid tone and/or endocannabinoid deficiency may be associated with a lessening of endocannabinoid-induced neuroprotective abilities—immune-modulating effects, pain reduction, and protection from oxidative stress, for example—and therefore may contribute to the biological pathologies of many a chronic condition.

THC and CBD

- THC and CBD are the prime constituents (cannabinoids) of cannabis that tend to occur in significant amounts to produce effects on the body, mind, and emotion. CBD and THC are also produced synthetically.
- THC is a multi-target molecule. Researchers have described more than 50 different pharmacological targets and effects of Δ9-THC.
- Similarly, CBD produces more than 30 pharmacologically recognized actions.
- **What They Do:** THC and CBD bind with a number of receptor sites that in turn

---

are responsible for a complex set of effects notable in body, mind, and emotion.

- **How They Do It:** Specifically, THC and CBD bind with ionotropic cannabinoid receptor sites, G-protein-coupled receptor sites, and peroxisome proliferator-activated receptors. Both also modulate specific ligands such as neurotransmitters, hormones, or signaling proteins. In addition, CBD produces effects by reducing the availability of enzymes that break down the body's own cannabinoids.

- **Why It Matters:** Scientific evidence clearly suggests that targeted use of both CBD and THC as well as in certain combinations can be used to prevent or mitigate a great number of conditions, especially many of those of a chronic or stubborn nature for which there is no cure in orthodox medicine.

- **THC Inhaling versus Ingesting:** Inhaled THC-containing remedies produce effects that come on quickly and are shorter lasting. Digested THC-containing cannabis products are longer lasting but come on much more slowly. When the effects set in, they can be sudden and strong.

- **The Five Primary Blocks of Receptor Sites/Mechanisms by Which THC and CBD Initiate Effects:**
  - Ionotropic cannabinoid receptor sites (TRPs)
  - G-protein-coupled receptor sites (CB1, CB2, GPCR 3, 6, 18, 55, and opioid receptors)
  - Ligand (e.g., neurotransmitters, hormones) receptor sites
  - Peroxisome proliferator-activated receptors (PPARs)
  - Endocannabinoid metabolizing enzymes

- **THC Dosage Guidelines (based on dosages used in the currently available human trials)**
  - THC micro dose: ~0.1 mg to 0.4 mg
  - THC low dose: ~0.5 mg to 5 mg
  - THC medium dose: ~6 mg to 20 mg
  - THC higher dose: ~21 mg to 50+ mg

- **CBD Dosage Guidelines (based on dosages used in the currently available human trials)**
  - CBD low dose: 0.4 mg to 19 mg
  - CBD medium dose: 20 mg to 99 mg
  - CBD high dose: 100 mg to 800+ mg

Chapter 4

Which Chemotype Is for You

In this chapter you will learn the following:
- All about cannabis chemotypes
- Who may benefit from a cannabis chemotype I, II, and III

As seen in previous chapters, the current cannabis landscape is still not welcoming in terms of information and access. Too much confusion and difficulty face new patients, experienced cannabis-using patients, and recreational users alike. In some cases, just knowing specific limitations or potential pitfalls helps you discern and prioritize information, allowing you to be an empowered cannabis patient, practitioner, user, or connoisseur. At this point in the evolution of the cannabinoid health sciences, it is up to each of us to learn what we can in order to make more informed decisions about cannabis use—for both medical treatment purposes and so-called recreational use (though one could make a case that all substance use is in some ways "medicinal").

And research points us toward the most empowering information, numbers you can use yourself. From now on we will be looking at solutions in a practical, step-by-step approach. This chapter introduces a way to think about cannabis that will allow you to break through the confusion and difficulty described earlier. When it comes to achieving and sustaining precise therapeutic effects, the three chemotypes of cannabis are an essential first step.

In a nutshell: Chemotype I contains more THC than CBD. Chemotype II contains relatively equal amounts of THC to CBD (close to 1:1). Chemotype III contains more CBD than THC (or no THC at all).[1] This ratio-based system was devised by the researchers E. Small and H. D. Beckstead in 1973,[2,3] and their method of classifying individual cannabis plants by THC and CBD content continues to inform the cannabinoid health sciences today. Every plant fits in one of the three chemotypes. You may wonder why simple ratio numbers are so important. These numbers determine not just what kind of cannabis experience you are going to have but also what therapeutic or adverse effects they are likely to produce. In other words, awareness of THC:CBD ratio numbers is the first step in a process that will allow you to dial in, produce, fine-tune, and sustain the precise effects you seek, in terms of symptom reduction and/or mood improvements.

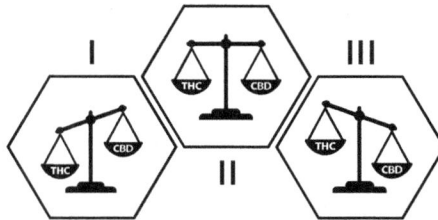

Chapter 4, Graphic 1: Cannabis Chemotypes

Who May Benefit from a Chemotype I?
(More THC than CBD)

Based on what you have read so far, you will have some basic understanding of the three cannabis chemotypes. Here, on these following few pages, we anchor what you already know and take a deeper look. Most commonly when people talk about their fears of using cannabis, such as fears of getting "high," fears of adverse effects, or fears about its addiction potential, they are referring to a chemotype I of cannabis. With this chemotype, the line between an adverse effect and a therapeutic one can be razor thin. Additionally, any addiction potential is clearly associated with THC.

However, the very abilities of THC to produce these effects, unwanted and feared by some, are also the very qualities that make this chemotype so therapeutic and useful for others.

For instance, you may recall from earlier sections of this book that CB1 receptor sites are abundantly present in the central nervous system (CNS). THC binds with these receptor sites and initiates a number of effects that are of particular relevance to patients with pathologies affecting the CNS (the brain and spinal cord). Thus patients with conditions such as traumatic tears in the CNS (e.g., injuries, stroke), degenerative neurological disorders (e.g., Alzheimer's), or specific types of pain such as pathological pain (e.g., amyotrophic lateral sclerosis), severe pain associated with HIV/AIDS, and nociceptive pain tend to all benefit from a chemotype I abundant in THC. Conditions where an overexcitability of specifically stressed tissues causes tension or spasms (e.g., spinal cord injuries) also benefit from the sedative aspect of this chemotype. A carefully dosed chemotype I can also positively improve a number of mood disorders such as depression or various forms of anxiety. However, it can't be said it often enough. Here with a cannabis chemotype I there is a fine line between a therapeutic effect and an adverse effect.

Mental and emotional effects from chemotype I may range widely—from the positive experiences of euphoria, deep relaxation, new ways of thinking, melting away of stress, enhanced creativity, satisfying laugher (sometimes uncontrollable), mirth, improvements in mood, or enhanced libido, to experiences of a negative nature such as tension, anxiety, panic, paranoia, or a sense of impending doom.

The reader is reminded that "feeling high"—experiencing altered consciousness or extraordinary states of awareness—may be considered an adverse effect by some. But other people seek out these same effects for therapeutic, creative, or transpersonal purposes. Most adverse effects are easily managed by reassurance and putting yourself in a safe, calm, relaxing atmosphere. Play gentle music or nature sounds, hold hands if desired, or just have the comforting presence of another person—it can make the difference.

Mitch's Story

Mitch had been a heroin addict for over a decade. For him, it all started with a motorcycle accident that severely injured his back. When he was released from the hospital, he was hooked on opioids. With little resources left and unable to work, he quickly turned to a cheaper version, heroin. When I first met him, he told me that: the past many years felt like living in a pain amplifier with no hope or an end in sight. My addiction was a war with everything, myself and especially the people who loved me. Somehow, heroin made it all OK. At least for a while. When the high wears off, all the pain, all the shame, the pity, the hopelessness comes flooding back, and with it you get the shakes, the spasms and cramps, the sweat and the stank. I was desperate, and in such a state, I don't even remember how I made it to the South of Market cannabis lounge, where a friend offered me a dab. I took one hit and the shakes were virtually gone. At first I couldn't believe it, but it worked. It worked without me passing out, it worked without a needle, and it didn't affect my breathing in any shape or form. In other words, it wasn't going to kill me, unlike the heroin, the only other thing I knew that worked so well in this situation."

In this case the sedative properties of large amounts of THC provided much-needed relief. However, keep in mind that if a person with a very different constitution, or not accustomed to the chronic use of strong drugs such as heroin, were to inhale the same dab, it could produce, near instantly, an undesirable experience such as that of Maureen Dowd described in the introduction.

Also, you may want to know that Mitch ultimately won his war with his addiction. His journey to health also serves as a reminder that the need for a specific chemotype can change in direct relationship to the healing one realizes. Case in point: Mitch moved slowly but surely down from using a concentrate (dab) to a chemotype I strain (Train Wreck) that continued to provide relief of his cramps, spasms, and dysphoria when needed. He eventually was able to switch over to a One to One strain (a chemotype II) maintaining great results with steadily lower amounts of THC. He noticed this trend and decided to continue in this direction of reducing THC and increasing the amount of CBD. It was not long until he was simply relying on a Charlotte's Web strain (a chemotype III) with very little THC at all. Now, to be fair, Mitch made it clear that cannabis was not the only reason that got him clean. He credits a combination of moving to a new place with safe access to medicinal grade "weed," a positive support system, a nutritional program, and a mindfulness practice (emotional self-regulation) he took up along the way. In fact, he has become so resilient that now even in the face of stress or in the presence of what would normally be a relapse trigger, that he hardly needs any cannabis-based support at all.

Cases such as Mitch's constantly remind me that an out-of-balance endocannabinoid tone or deficiencies can be corrected by the careful concomitant use of plant-based cannabinoids as well as by endocannabinoid balancing inner work (e.g., mindfulness).

CB1 activation via THC also suppresses nausea and vomiting, and thus patient populations that benefit from a chemotype I include cancer patients, especially those undergoing chemotherapy and experiencing otherwise uncontrollable nausea and vomiting—while also suffering from stress, tension, insomnia, pain, nephrotoxicity, cardiotoxicity, and a general loss of quality of life. An appropriately dosed use of a chemotype I can make a significant difference in mitigating all primary and secondary symptoms.

Further, THC reduces cough responses, functions as a bronchodilator (opens the airways), and through binding with CB2 receptor sites, which are abundantly present in the body's immune system, where it, when activated, downregulates immune functions. Thus patients suffering from an overactive immune system (e.g., allergic asthma, chronic coughing) often find relief in strains belonging to this chemotype.

A pharmaceutical version of a chemotype I is the FDA-approved, synthetic drug Dronabinol (Marinol, Syndros). It contains a THC:CBD ratio of 1:0. Treatment examples for Dronabinol include appetite stimulation to reduce weight loss in patients with AIDS, and it is used as an anti-emetic in cancer patients experiencing chemotherapy-induced nausea and vomiting. Potential adverse effects of Dronabinol and chemotype I cannabis may include "feeling high," altered consciousness, paranoia, anxiety, insomnia, pain, or nausea.

The following summary of chemotype I strains and uses (plus those for chemotypes II and III) is far from exhaustive. Note that if you are not drawn to pharmaceutical cannabis compounds such as Dronabinol, your process of finding a similar ratio and cannabis effect will be more complex, though ideally it leads to a much more personalized form of medicine and you will get the added benefits of the other plant constituents that tend to harmonize and even amplify therapeutic effects.

Chemotype I Summary

Top three strain examples

- Godfather OG (~34% THC)[4]
- Super Glue (~32% THC)[5]
- Strawberry Banana (~32% THC)[6]

Pharmaceutical versions:

- Dronabinol
- Marinol
- Syndros

When selecting a chemotype I, my priorities are likely to include the following:

- I am not concerned about getting "high"
- A higher risk for adverse effects is acceptable
- I want or am OK with changes in cognition
- I like big changes in mood improvements (remember there is a fine line between a therapeutic or adverse effect)
- I want sedative effects but remain aware (conscious sedation)
- I need deep relaxation in body, mind, and emotion

Sample effects with relevance to a chemotype I include the following:

Mental/Emotional Effects

Cognition

- "Feeling high," ranging from a pleasant euphoria to an unpleasant self-consciousness/paranoia/panic attack

Memory

- Increases the ability to relax and be in the presence of intensely fearful memories, a positive effect in treating certain anxieties or phobias
- Slows habitual memory, helpful in cases of OCD and addiction triggers, for example
- Impairs short-term and long-term memories (normally THC-dose-dependent and temporary)

Mood

- Deep relaxation and stress reduction
- Conscious sedation (especially therapeutic in overly excited/stressed nervous system-based conditions; see sample conditions below)
- Significant mood improvement is possible, but there is a very narrow line between anxiolytic and anxiogenic effects, ranging from euphoria (joy) to dysphoria (feeling anxious, scared, or panicky)

Physiological Effects

- Bronchodilation (e.g., allergic asthma, cough)
- Appetite stimulation (weight loss, anorexia, cachexia)
- Immune-system modulation (down-regulation is relevant to conditions in which the body's immune system begins to attack itself, e.g., rheumatoid arthritis)

Pain types that may benefit

- Central pain (e.g., stroke)
- Pathological pain (e.g., amyotrophic lateral sclerosis)
- Nociceptive pain (noxious stimulus from e.g., chemical, heat, mechanical sources)
- Mental-emotional pain

Sample conditions that may benefit from a chemotype I include the following:

- HIV/AIDS-based neuropathies
- Severe physical withdrawal symptoms (e.g., heroin, opioids, methamphetamines)
- Chemotherapy-induced nausea and vomiting
- Spinal cord injury–induced chronic muscle spasms
- Alzheimer's disease with nighttime agitation or disturbed behavior
- Autism with agitation with self/other-harming behavior

Addiction Potential: Low but there is a potential risk for psychological addiction.

Cannabis Chemotype I:
(effects on the left are amplified while those on the right, in gray, are diminished)

THC		

Receptors	Response	**Effects**

TRPs		
TRPA1	△ Agonist[2]	May ↓ pain, itch, & obesity[1]
TRPV1	N/A	N/A
TRPV2	△ Agonist[4]	May ↓ heat, pain, inflammation[3]
TRPV4	N/A	N/A
TRPM8	▽ Antagonist[6]	May ↓ pain[5]

GPCR		
Opioid γ/δ	(+)Allo. mod.	Synergistic analgesia (6×<CBD)[7]
GPCR **CB1**	△ Agonist	CNS (e.g., ↑ ↓ cognition/mood)
GPCR **CB2**	△ Agonist	↓ Immune responses,[8] ↑ ↓ mood[9]
GPCR 3	N/A	N/A
GPCR 6	N/A	N/A
GPCR 18	△ Agonist	Testes, spleen,[10] may ↓ high BP[11]
GPCR 55	△ Agonist[17]	Heart funct.,[13] may ↑ risk for cancer[14–16]

Ligands	(+/-) Allo. mod.	↑positive affect/mood

PPAR's γ	△ Agonist γ[20,21]	May ↓ hypertension,[18] ↓ cancer[19]

Enzymes	N/A	N/A

N/A not applicable
↑ Increases
↓ Decreases
△ Agonist (produces action)
▽ Antagonist (weakens/blocks action)　　☐ amplified　　☐ diminished

CBD

Receptors	Response	Effects
TRPs		
TRPA1	$\triangle$ Agonist[22]	May $\downarrow$ pain, itch, & inflammation
TRPV1	$\triangle$ Agonist[23,24]	May $\downarrow$ pain[25]
TRPV2	$\triangle$ Agonist[26]	May $\downarrow$ chronic pain[27]
TRPV4	$\triangle$ Agonist[28]	May $\downarrow$ acne[29]
TRPM8	∇ Antagonist[30]	$\downarrow$ Inflammation,[31] $\downarrow$ prostate ca.[32]
GPCR		
Opioid γ/δ	(+)Allo. mod.	$\uparrow$ Synergistic analgesia (6×>THC)[33]
GPCR **CB1**	$\triangle$ Ago.[34] ∇ Anta.*[37]	CBD $\downarrow$ effect of THC[35,36]
GPCR **CB2**	$\triangle$ Ago.[38] ∇ Anta.*[41]	$\downarrow\uparrow$ Immune,[39] $\uparrow$ mood[40]
GPCR 3	$\Diamond$ Inverse agonist[42]	$\downarrow$ Alzheimer's,[43] neuropathy[44]
GPCR 6	$\Diamond$ Inverse agonist[45]	$\downarrow$ Parkinson's disease[46]
GPCR 18	N/A	N/A
GPCR 55	∇ Antagonist[47]	$\downarrow$ Bone loss,[48] may $\downarrow$ certain cancers[49–51]
Ligands	(+/-) Allo. mod.	e.g., $\uparrow$ 5HT1A[52] $\uparrow$ GABA$_A$[53] $\downarrow$ Adenosine[54]
PPAR's γ	$\triangle$ Agonist [55]	$\downarrow$ Lung ca.,[56] $\downarrow$ inflammation & pain[57,58]
Enzymes	$\downarrow$FAAH[59] $\uparrow$ AEA	$\uparrow$ Mood & $\downarrow$Psychosis[60]

$\Diamond$ Inverse agonist (opposite action)
(+) Positive allosteric modulator (amplifies action of an agonist)
(-) Negative allosteric modulator (diminishes action of an agonist)
δ-delta α-alpha, μ-mu, γ-gamma
*Selective allosteric antagonist of CB1/CB2 receptor agonists

Chapter 4, Graphic 2: Cannabis Chemotype I

Who May Benefit from a Chemotype II?
(Relatively Equal THC and CBD)

One of the general differences between a chemotype I strain and a chemotype II strain is that THC tends to support the therapeutic abilities of CBD, while CBD tends to tame the potential adverse effects associated with THC. A chemotype II gives certain patients their necessary central nervous system CNS activation, but in a more tempered way than a chemotype I. Thus using a chemotype II reduces the risk of changes in cognition (getting "high") and the potential for any adverse effects, and further lowers the risk of developing addiction.

Mental and emotional effects can differ widely from those of chemotype I cannabis strains. Many patients who use chemotype II describe the differences in terms of medium relaxation with a lighter sensation of feeling high accompanied by a sense of uplift in mood, moderate euphoria, an easy smile, a calmness of spirit, serenity, or tranquility. Fewer adverse effects occur with chemotype II. Some more sensitive patients, especially when using higher dosages of THC, may experience the potential adverse effects, but if they do, effects are usually milder and shorter in duration.

If you or a patient in your care could benefit from any of the effects or treatment examples listed in the chemotype I section but is very concerned about the potential risks, a chemotype II may be a better choice. You still get the activation of THC at CB1 and CB2 but at a lower potency. Here, in comparison to the use of a chemotype I, the line between a therapeutic and an adverse effect is much wider, providing you with a significant buffer from any potential ill effects that are THC-dose-dependent.

Here are a couple of practical examples of using a chemotype II. A number of clinical trials have shown that a chemotype II was effective in treating primary and secondary symptoms of multiple sclerosis such as reduction of muscle spasms, pain related to spasms, or frequency of bladder incontinence. For treating various types of pain, a chemotype II trend has emerged, suggesting its use for certain peripheral neuropathies associated with cancer, cancer treatments (e.g., cisplatin), or auto-immune disorders (e.g., multiple sclerosis). This is especially important in the context of treating chronic neuropathies where the use of opioids has demonstrated poor performance results.[7-9]

One of the few double-blind, placebo-controlled human trials that compared the efficacy of all three cannabis chemotypes side by side in the treatment of a single condition (fibromyalgia) comes to us from the Netherlands, where researchers discovered that a single inhalation (using a vaporizer) of a chemotype II (Bediol® 6.3% THC : 8% CBD) with the ratio of ~1:1.3 (THC:CBD) produced a 30% reduction in pain.

Of late, the academic landscape has been very excited about the scientific findings suggesting efficacy of cannabis in treating various cancers. The emerging picture is very complex but already suggests some more considerations. Here is one example in the context of choosing the most appropriate chemotype. While THC has been shown to produce a number of pathways by which it creates apoptotic (anti-cancer) effects, a number of pre-clinical trials have discovered that colon, ovarian, and metastasizing breast cancer cell lines rely, at least in part, on GPCR-55 for proliferation.[10-12] Since THC is an agonist at GPCR 55 and CBD an antagonist at the same receptor, these early study results suggest caution in using a chemotype I for these particular cancers. While it might turn out that THC's other than GPCR-55-based anti-cancer effects outweigh the potential pitfalls of GPCR-55 activation on possible proliferation, until the science is more compelling in these specific circumstances, a chemotype II concentrate may be a more appropriate option.

A pharmaceutical example of a chemotype II is a class of drugs called Nabiximols such as Sativex. It is a fully FDA-approved drug made from whole plant cannabis, including a number of other plant constituents, to treat two pediatric seizure disorders: Dravet syndrome and Lennox-Gestaut syndrome. In a number of other countries, it is also approved as an additional pain medication for cancer patients who find no relief even with the highest allowable dose of opioids.[13] Adverse effects of Sativex may include dizziness or fatigue (very common); dry mouth, disorientation, depression (common); and throat irritation/application irritation, feeling very high, rapid heart rate (uncommon).[14]

And, as you might suspect, adverse effects of chemotype II strains mirror those of Sativex and may include dizziness, dry mucous membranes (mouth), and disorientation.

Chemotype II Summary
Top three strain examples
- Cellar CBD, ~7.3% THC to ~7.7% CBD[15]
- Nubia, ~8% THC to ~7.5% CBD[16]
- Hayley's Comet, with THC less than 15% and CBD greater than 10%, comes close[17]

Pharmaceutical versions
- Nabiximols (e.g., Sativex) with a THC:CBD ratio of ~1:1

When selecting a chemotype II, my priorities are likely to include the following:
- I am moderately concerned about getting "high"
- A medium risk for adverse effects is acceptable
- I am OK with the potential for some changes in cognition (dose dependent on actual THC amount used)
- I want moderate changes in mood improvements
- I need moderate relaxation in body, mind, and emotion

Sample effects with relevance to a chemotype II include the following:
Mental/Emotional Effects
Cognition
- Lower risk of feeling "high" but still some potential for changes in cognition (THC-dose-dependent)

Memory
- Similar changes to various types of memory functions as when using a chemotype I but much milder in nature
- Increased CBD content reduces THC-induced memory impairments

Mood

- Moderate relaxation and stress reduction
- Low to moderate sedative affects (therapeutic in moderately overexcited/stressed nervous system–based conditions)
- Mood improvement (wider line between anxiolytic and anxiogenic effects)
- Moderate effects as anxiolytic or anxiogenic

Physiological Effects

Pain types that may benefit

- Peripheral neuropathies
- Pain due to muscle spasms (e.g., multiple sclerosis)
- Cancer treatment-induced or cancer-based neuropathies
- Mental-emotional pain

Sample conditions that may benefit from a chemotype II include the following:

- Multiple sclerosis (calms and reduces pain, spasm, urinary incontinence)
- Fibromyalgia (provides systemic analgesia)

Addiction Potential: There is a very low risk of developing addiction (THC-dose-dependent)

Judy's Story

One of the earlier indications that a chemotype II may provide a number of benefits to patients suffering from multiple sclerosis came from a a study titled "Standardized Cannabis in Multiple Sclerosis: A Case Report."[18] In MS the body's immune system begins to attack the body's own cells; more specifically, those cells that insulate nerve fibers (myelin sheets). When the nerves are progressively laid bare, nerve conduction is impaired, leading to a number of classical symptoms such as tingling and numbness, twitching and spastic activities, pain (often like electricity), difficulty maintaining bladder control (incontinence), unsteady gait, or slurred speech. Within the orthodox medical system its exact cause is poorly understood, and to date, there is no cure for this condition, highlighting the need for novel or innovative approaches.

Judy (fictitious name) was a 52-year-old woman suffering from MS. She was the focus of a case study conducted by the Green Cross Society of British Columbia in Canada. The medical team followed her progress for a year where she was given a standardized amount of cannabis. Upon initial contact with the society, she suffered from chronic pain (especially severe in her left foot), muscle tremors, and an unsteady gait. She received a tested, standardized edible cannabis-containing product abundant in THC, CBD, and CBN. Her caregiver provided daily progress reports, charting a number of details such as quality (e.g., burning, stabbing, shooting, aching, numbness, or cramping) and intensity of pain (on a scale from 1 to 10, 10 being the worst). The reports were assessed daily through verbal dialogue with the caregiver (with a focus on the amount of daily THC). Results showed an immediate improvement after using the oral preparation. Pain scores went from an 8-10 to 1-2 over the first month, during which an optimal dosing regimen was established. The study reported that the greatest benefit for pain and tremor was achieved consuming 6-8, 50 mg capsules per day, dosing at four hour intervals. Judy also consumed about four to six cannabis-only cigarettes in case she needed needed help with breakthrough pain or general improvements in mood. Measurement of the average THC content was equal to about 25 mg per cigarette, putting her daily amount of THC to 400 to 500 mg. Judy continued to improve from virtual incapacitation to where she was able to dress herself, go for walks, and even do some gardening during the year. Over time dosage and ratios were continuously monitored and adjusted. Eventually, it was found that Judy experienced optimal relief with strains containing relatively high CBD, indicating a beneficial trend moving from a chemotype I to a chemotype II.

Cannabis Chemotype II
(effects on both sides express relatively equally)

THC		

Receptors	Response	Effects

TRPs		
TRPA1	△ Agonist[2]	May ↓ pain, itch, & obesity[1]
TRPV1	N/A	N/A
TRPV2	△ Agonist[4]	May ↓ heat, pain, inflammation[3]
TRPV4	N/A	N/A
TRPM8	▽ Antagonist[6]	May ↓ pain[5]

GPCR		
Opioid γ/δ	(+)Allo. mod.	Synergistic analgesia (6×<CBD)[7]
GPCR **CB1**	△ Agonist	CNS (e.g., ↑ ↓ cognition/mood)
GPCR **CB2**	△ Agonist	↓ Immune responses,[8] ↑ ↓ mood[9]
GPCR 3	N/A	N/A
GPCR 6	N/A	N/A
GPCR 18	△ Agonist	Testes, spleen,[10] may ↓ high BP[11]
GPCR 55	△ Agonist[17]	Heart funct.,[13] may ↑ risk for cancer.[14-16]

Ligands	(+/-) Allo. mod.	↑ positive affect/mood

PPAR's γ	△ Agonist μ[20,21]	May ↓ hypertension,[18] ↓ cancer[19]

Enzymes	N/A	N/A

N/A not applicable

↑ Increases

↓ Decreases

△ Agonist (produces action)

▽ Antagonist (weakens/blocks action)

CBD

Receptors	Response	Effects
TRPs		
TRPA1	△Agonist[22]	May ↓ pain, itch, & inflammation
TRPV1	△Agonist[23,24]	May ↓ pain[25]
TRPV2	△Agonist[26]	May ↓ chronic pain[27]
TRPV4	△Agonist[28]	May ↓ acne[29]
TRPM8	▽Antagonist[30]	↓ Inflammation,[31] ↓ prostate ca.[32]

Receptors	Response	Effects
GPCR		
Opioid γ/δ	(+)Allo. mod.	↑ Synergistic analgesia (6×>THC)[33]
GPCR **CB1**	△Ago.[34] ▽Anta.*[37]	CBD ↓ effect of THC[35,36]
GPCR **CB2**	△Ago.[38] ▽Anta.*[41]	↓ ↑ Immune,[39] ↑ mood[40]
GPCR 3	◊Inverse agonist[42]	↓ Alzheimer's,[43] neuropathy[44]
GPCR 6	◊Inverse agonist[45]	↓ Parkinson's disease[46]
GPCR 18	N/A	N/A
GPCR 55	▽Antagonist[47]	↓ Bone loss,[48] may ↓ certain cancers[49-51]

Ligands	(+/-) Allo. mod.	e.g., ↑ 5HT1A[52] ↑ $GABA_A$[53] ↓ Adenosine[54]
PPAR's γ	△Agonist [55]	↓ Lung ca.,[56] ↓ inflammation & pain[57,58]
Enzymes	↓FAAH[59] ↑AEA	↑ Mood & ↓ Psychosis[60]

◊ Inverse agonist (opposite action)
(+) Positive allosteric modulator (amplifies action of an agonist)
(-) Negative allosteric modulator (diminishes action of an agonist)
δ-delta α-alpha, µ-mu, γ-gamma
*Selective allosteric antagonist of CB1/CB2 receptor agonists

Chapter 4, Graphic 3: Cannabis Chemotype II

Who May Benefit from a Chemotype III?
(More CBD than THC, or No THC)

If you are a patient or are taking care of someone who wants to make sure there is no risk of getting "high" (changes in cognition), if you want to avoid any of the adverse effects commonly associated with cannabis use in general, and be sure that there is not even a small risk of becoming addicted, this chemotype is for you (e.g., Cannatonic, AC/DC). Mental and emotional effects include a gentle uplift in mood with no effects on cognitive abilities. Thinking, perception, sensory experiences remain unaltered. However, there is one caveat, in some very sensitive patients, a chemotype III that contains a still-significant level of THC (e.g., CBD OG) may still produce adverse effects such as anxiety, although in much milder forms compared to high-THC chemotype I. This possibility is easily shored up by using strains with no or only a very low amount of THC (<0.3%) and that have been tested by a reputable laboratory. Knowing the numbers of THC present in your chemotype III will give you agency and confidence about what kind of cannabis experience you wish to have. Take a look at some examples of common chemotype III strains: Cannatonic, 0.8% THC:20.8% CBD[19]; CBD OG, 9.01% THC : 19.5% CBD[20]; or AC/DC, 0.8% THC:15.4% CBD.[21]

Strains abundant in CBD and with only small or trace amounts of THC are rich, in their own right, in being able to produce a great number of proven therapeutic benefits. For instance, patients in need of a general uplift in mood tend to benefit from a chemotype III. Patients with a number of chronic degenerative neurological illnesses with underlying or contributing pathologies such as oxidative stress and neurodegeneration such as seizure disorders (especially pediatric seizures), drug dependencies (e.g., alcohol, opioid, methamphetamines)—or those dealing with mental conditions such as depression, anxiety, or psychotic disorders—may seek a chemotype III medicine. Patients with serious pain who rely on opioid treatment often utilize a chemotype III to create a synergistic effect; it lets them reduce the opioids while achieving the same analgesic effect, with the benefit of lower costs and a reduction in dependence and overdose risks.

A pharmaceutical version of a chemotype III includes the FDA-approved drug Epidiolex, a plant-derived cannabidiol-only medication.[22] Treatment examples from various countries where Epidiolex is legal include treatment-resistant epilepsy syndromes such as Dravet syndrome, Lennox-Gastaut syndrome, Tuberous Sclerosis complex, and infantile spasms.[23] Adverse effects of Epidiolex most commonly reported (from among severely sick patients) may include drowsiness (somnolence) and changes in appetite and weight (up or down). A study conducted on pediatric patients suffering from treatment-resistant epilepsy demonstrated that Epidiolex is a promising treatment for these children, and it is generally well-tolerated in doses up to 25 mg/kg/d.[24]

Chemotype III strains with no or only trace amounts of THC are generally considered safe for humans, including for chronic use and in high doses up to 1,500 mg/day.[25] However, some studies suggest that all chemotypes of cannabis including a chemotype III may interfere with certain pharmaceutical drugs. For instance, the concomitant use of clobazam (a pharmaceutical drug to treat seizures) with CBD makes the pharma drug more potent, leading to an increase in adverse effect potential such as drowsiness, ataxia, or irritability.[26] However, these adverse effects can be anticipated and avoided by lowering either clobazam or CBD amounts. The same concern applies to all other pharmaceutical drugs that are metabolized (broken down) by the enzymes belonging to the family cytochrome P450. This is because CBD inhibits the breakdown of these enzymes.

Furthermore, a chemotype III, in reaching for balance (homeostasis), may produce temporary tissue-specific pro- and/or anti-inflammatory responses, and thus can reduce inflammation in many patients or increase inflammation in certain especially severely immunocompromised patients such as those suffering from allergic inflammations of the lung.[27]

The reader may also want to take notice that purified, single-molecule CBD products may have limited therapeutic uses. Research conducted on mice in the context of treating inflammatory or nociceptive types of pain demonstrated superior effects of a full-spectrum (whole plant) CBD-abundant cannabis extract.[28]

Charlotte's Web

Once upon a time there was a little girl with the name of Charlotte Figi. Charlotte was born in 2006 and suffered from Dravet syndrome, a debilitating form of childhood seizures that is often very difficult to treat with the normal anti-seizure regimen of pharmaceutical drugs such as clobazam or Stiripentol. Once a seizure was triggered, it could last for hours and be considered a true emergency requiring rapid transport and treatment with regular visits by the paramedics and trips to the emergency room. After trying everything the orthodox medical system had in its repertoire and getting no or only very limited results, Charlotte's dad, a Green Beret serving in Afghanistan, suggested trying something new: cannabis. After doing a bit of research, the Figis got in touch with the Realm of Caring Foundation that is dedicated to bettering the lives of people affected with chronic debilitating conditions, and a partnership was born that changed the Figis' lives forever. They methodically tried out a number of different chemotypes of cannabis to see which one might result in providing much-needed relief. It turned out that a chemotype III was optimal and worked wonders for Charlotte. Her case was so obviously dramatic that it became a force to be reckoned with. Her story gained worldwide attention and was the impetus for a number of policy changes in the US needed to maintain an adequate supply of consistent plant constituents needed for optimal results. In honor of her story the strain is called Charlotte's Web. Her results have now been replicated by patients suffering from Dravet all around the globe. Paige Figi (Charlotte's mother) said of the strain named after her daughter: "It's helped everything. She has over 99% seizure control. She doesn't use her feeding tube anymore (she used gastric tube for eating before). She doesn't have her autistic behavior anymore, and she doesn't have severe sleep disorders." "She can walk—she's not in her wheelchair at all—and she's talking. She couldn't talk before, and now she's talking. It's been a totally life-changing event, totally life-changing medicine." Charlotte Figi passed away in April 2020. Her parents, Paige and Stephen Figi, believe that the cause of her death is related to complications from COVID-19.

Chemotype III Summary

Top strain examples

- Cannatonic (e.g., 0.8% THC:20.8% CBD)
- CBD OG (e.g., 9.01% THC:19.5% CBD)
- AC/DC (0.8% THC:15.4% CBD)
- Charlotte's Web (0.5% THC:17% CBD)

Pharmaceutical version

- Epidiolex THC:CBD is 0:1

When selecting a chemotype III, my priorities are likely to include the following:

- I am very concerned about getting "high"
- I want to avoid adverse effects
- I do not want any changes in cognition
- I want gentle and positive changes in mood improvements
- I need gentle relaxation in body, mind, and emotion

Sample effects with relevance to a chemotype III include the following:

Mental/ Emotional Effects

Cognition

- No risk of getting "high," and no changes in cognition (unless your strain contains more than 0.3% THC)

Memory

- Potential neuroprotection in cases of dementia improves memory function

Mood

- Mild relaxation and stress reduction; gentle uplift in affect (emotions)
- No sedative effects
- Mood improvements (anxiolytic with no anxiogenic effect if THC is less than 0.3%)

Physiological Effects

Pain types that may benefit

- General analgesic properties—possibly synergistic with opioids
- Inflammatory pain
- Chronic pain
- Mental-emotional pain

Sample conditions that may benefit from a chemotype III include the following:

- Acne
- Heart disease
- Colitis
- Mental or mood disorders
  - Depression
  - Psychosis

- ◆ Addiction
- ◆ Anxiety
- ◆ Post-traumatic stress disorder (PTSD)
- ◆ OCD
- Seizure disorders
  - ◆ Epilepsy (pediatric and adult)
  - ◆ Dravet syndrome
  - ◆ Lennox-Gastaut syndrome
  - ◆ Tuberous Sclerosis complex
  - ◆ Infantile spasms

Addiction Potential: There is none if the THC amount is less than 0.3%.

The treatment indications as well as adverse effects for the pharmaceutical versions and corresponding strains tend to mirror each other.

Two additional chemotypes have been proposed in the scientific literature: chemotypes IV and V. Neither contributes unique, practical, or therapeutically significant information not already covered by chemotypes I through III, or by reports on the biological impact of individual cannabinoids. Chemotype IV reported by Fournier et al. in 1987[29] is defined as having CBG as the most abundant cannabinoid. Chemotype V proposed by Mandolino and Carboni in 2004[30] includes strains that are void of any detectable cannabinoids. Keep in mind that naming taxonomies are meant to change as new information becomes available. For instance, given emerging relevancies of THCV as a potentially effective treatment for patients with metabolic syndrome, diabetes, or obesity I wouldn't be surprised if eventually the governing naming institutions were to designate additional chemotypes such as a chemotype abundant in THCV.

Key Points, Chapter 4: Which Chemotype Is for You

- Of all cannabinoids, THC and CBD tend to have the greatest relevance to most people.
- Therapeutic capacities of a type of cannabis plant are significantly determined by the ratio of THC to CBD.
- THC is the primary mind-altering cannabinoid. CBD produces affectional changes.
- To minimize potential adverse effects, consider using the lowest THC amount possible to still obtain desired effects. CBD often produces similar effects, but by different mechanisms. Thus you may benefit from cannabis flower or products primarily or solely using CBD.

Cannabis Chemotype III
(effects on the right are amplified
while those on the left are diminished)

THC

Receptors	Response	Effects
TRPs		
TRPA1	△ Agonist[2]	May ↓ pain, itch, & obesity[1]
TRPV1	N/A	N/A
TRPV2	△ Agonist[4]	May ↓ heat, pain, inflammation[3]
TRPV4	N/A	N/A
TRPM8	▽ Antagonist[6]	May ↓ pain[5]
GPCR		
Opioid γ/δ	(+)Allo. mod.	Synergistic analgesia (6×<CBD)[7]
GPCR **CB1**	△ Agonist	CNS (e.g., ↑↓ cognition/mood)
GPCR **CB2**	△ Agonist	↓ Immune responses,[8] ↑↓ mood[9]
GPCR 3	N/A	N/A
GPCR 6	N/A	N/A
GPCR 18	△ Agonist	Testes, spleen,[10] may ↓ high BP[11]
GPCR 55	△ Agonist[17]	Heart funct.,[13] may ↑ risk for cancer.[14–16]
Ligands	(+/-) Allo. mod.	↑ positive affect/mood
PPAR's γ	△ Agonist γ[20,21]	May ↓ hypertension,[18] ↓ cancer[19]
Enzymes	N/A	N/A

N/A not applicable
↑ Increases
↓ Decreases
△ Agonist (produces action)
▽ Antagonist (weakens/blocks action)

[ ] diminished   [ ] amplified

CBD

Receptors	Response	Effects
TRPs		
TRPA1	△ Agonist[22]	May ↓ pain, itch, & inflammation
TRPV1	△ Agonist[23,24]	May ↓ pain[25]
TRPV2	△ Agonist[26]	May ↓ chronic pain[27]
TRPV4	△ Agonist[28]	May ↓ acne[29]
TRPM8	▽ Antagonist[30]	↓ Inflammation,[31] ↓ prostate ca.[32]
GPCR		
Opioid γ/δ	(+)Allo. mod.	↑ Synergistic analgesia (6×>THC)[33]
GPCR **CB1**	△ Ago.[34] ▽ Anta.*[37]	CBD↓effect of THC[35,36]
GPCR **CB2**	△ Ago.[38] ▽ Anta.*[41]	↓ ↑ Immune,[39] ↑ mood[40]
GPCR 3	◇ Inverse agonist[42]	↓ Alzheimer's,[43] neuropathy[44]
GPCR 6	◇ Inverse agonist[45]	↓ Parkinson's disease[46]
GPCR 18	N/A	N/A
GPCR 55	▽ Antagonist[47]	↓ Bone loss,[48] may ↓ certain cancers[49–51]
Ligands	(+/-) Allo. mod.	e.g., ↑ 5HT1A[52] ↑ GABA$_A$[53] ↓ Adenosine[54]
PPAR's γ	△ Agonist [55]	↓ Lung ca.,[56] ↓ inflammation & pain[57,58]
Enzymes	↓FAAH[59] ↑AEA	↑ Mood & ↓ Psychosis[60]

◇ Inverse agonist (opposite action)
(+) Positive allosteric modulator (amplifies action of an agonist)
(-) Negative allosteric modulator (diminishes action of an agonist)
δ-delta α-alpha, μ-mu, γ-gamma
*Selective allosteric antagonist of CB1/CB2 receptor agonists

Chapter 4, Graphic 4: Chemotype III

Chapter 5

The CannaKeys Process

In this chapter you will learn the following:
- How to dial in and sustain desired effects (Step 1-6 of the CannaKeys Process)
  - ◆ Discover your specific chemotype
  - ◆ Narrow down your ideal ratio of THC:CBD
  - ◆ Choose your optimal form
  - ◆ Determine your proper dose
  - ◆ Design the best possible entourage effect for you
  - ◆ Learn how to keep a session log to fine-tune desired effects

In my years of doing research and writing in and about the cannabinoid health sciences, I have come to believe that the reason why cannabis works for so many different types of patient populations is that it affects the body and the mind in such a way as to support and fortify the body's own capacity for self-healing. A properly chosen and dosed cannabis experience matched to the right patient does so in three different ways: it produces a deep sense of relaxation; it creates a propensity for laughter or lightheartedness; and it induces multiple system homeostasis at a cellular level.

Turning cannabis into a precision medicine that consistently delivers the exact effect you want is the purpose of this chapter. It may appear like a tall order or a very complex undertaking. However, it is this very complexity that makes it easier to match the right type of cannabis to meet your needs and preferences. It is here you will realize that you have agency about what kind of cannabis experience you wish to have. With each completed step you can make more informed decisions and take evidence-based actions on your path to optimal health and well-being.

To be sure, in working with cannabis there are a number of variables that are beyond your control, such as the speed of cannabinoid delivery to the brain (movement across the blood brain barrier), the specific amounts of active ingredients lost in vapor, or the mg amount of THC in an edible that is converted by the liver into metabolites. However, there are six active measures you can take to ensure precision effect and consistent results. These variables are subject to conscious control and serve as a step-by-step process to find the medicine that produces optimal results for you. We have named this course of action the CannaKeys Process.

Step 1: Discovering Your Specific Chemotype of Cannabis

THC and CBD, the most abundant and biologically active of all cannabis constituents, play a significant role in the effects cannabis can produce. We use a THC-to-CBD ratio system to identify individual cannabis plants as one of three discrete chemotypes, I, II, and III. Discovering your ideal type is the first step to turn your cannabis prescription into precision medicine.

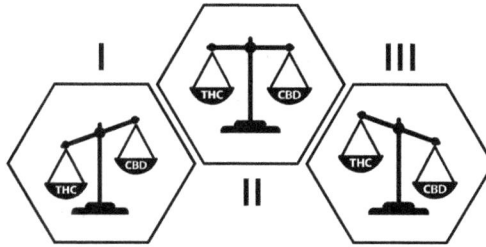

Chapter 5, Graphic 1: Discovering Your Cannabis Chemotype

By understanding the broad effects that each of these three types of cannabis can produce, you take an initial degree of control by expressing your preference. Once more, remember the story of Maureen Dowd from *The New York Times* who wrote about her misadventures with an edible. If Dowd had been aware that a chemotype I cannabis-based product contains a high amount of psychoactive THC—and chemotype II contains significantly less (additionally tamed by CBD), and chemotype III little or no THC—she would have been able to choose the degree or total lack of psychoactivity of her experience.

Choosing the degree of psychoactivity is only one element that empowers you as a patient. Another factor is that one of the three types may be better suited to your conditions or symptoms. When it comes to mood improvements, alleviating different kinds of pain, and producing therapeutic effects in a number of different chronic conditions, one cannabis chemotype may work better than another. For instance, a chemotype II tends to work well for many patients suffering from primary and secondary symptoms of MS, a chemotype I tends to work well for patients suffering from central pain associated with ALS or chronic spasms common in spinal cord injuries, and a chemotype III has served a number of patient populations suffering from various seizure disorders (especially pediatric fits unresponsive to pharmaceutical medication) and psychiatric conditions (anxiety or depression).

Step 1 starts with marking your priorities (about what kind of cannabis experience you wish to have), effects you wish to realize, and whenever possible, indicated conditions. Once you fill out the following form, use the emerging cluster (or pattern) as a first point of discernment in choosing the type that is likely to work best to meet your present needs.

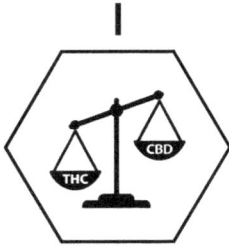

Your priorities when choosing a chemotype I

☐   I am not concerned about getting potential cognitive changes

☐   A higher risk for adverse effects is acceptable

☐   I want changes in cognition and consciousness

☐   I need deep relaxation in body, mind, and emotion

☐   I want big changes in mood improvements

☐   I need a sedative effect

☐   I am OK with the small risk of psychological addiction

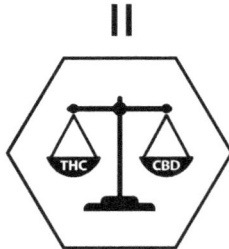

Your priorities when choosing a chemotype II

☐   I am moderately concerned about getting potential cognitive changes

☐   A lower risk for adverse effects is acceptable

☐   I am OK with minor changes in cognition and consciousness

☐   I want moderate changes in mood improvements

☐   I need moderate relaxation in body, mind, and emotion

☐   I like to employ a milder sedative effect

☐   I am OK with the very small risk of addiction

Common priorities when choosing a chemotype III

☐   I am very concerned about getting potential cognitive changes

☐   I don't want any changes in cognition

☐   I want gentle changes in mood improvements

☐   I need gentle relaxation in body, mind, and emotion

☐   I want to avoid any sedative effect

☐   I want to avoid serious adverse effects and any addiction potential

Additionally, you may find the following evidence-based chart that includes common concerns about mood improvements, analgesia for a specific type of pain, and some sample conditions helpful as a starting point. These emerging trends are limited by what has been studied to date, but they present a point of discernment to make more informed decisions. To list all of the emerging scientific trends and the chronic patient populations that may benefit from knowing them is beyond the scope of this book. Simply select the relevant boxes in the section below to see which ratio lights up for you here.

Sample Effects with Relevance to Chemotype I

Mental/Emotional Effects

Cognition

☐ "Feeling high," ranging from a pleasant euphoria to anxiety

Memory

☐ Increases the ability to relax and be in the presence of intensely fearful memories, a positive effect in treating certain anxieties or phobias

☐ Slows habitual memory, may be helpful in cases of OCD and addiction triggers, for example

☐ Impairs short-term and long-term memories (normally THC-dose-dependent and temporary)

Mood

☐ Deep relaxation and stress reduction

☐ Conscious sedation (especially therapeutic in overly excited/stressed nervous system–based conditions; see sample conditions below)

☐ Significant mood improvement is possible, but there is a very narrow line between anxiolytic and anxiogenic effects, ranging from euphoria (joy) to dysphoria (feeling anxious, scared, or panicky)

Physiological Effects

☐ Bronchodilation (e.g., allergic asthma, cough)

☐ Appetite stimulation (weight loss, anorexia, cachexia)

☐ Immune-system modulation (down-regulation is relevant to conditions in which the body's immune system begins to attack itself, e.g., rheumatoid arthritis)

Pain types that may benefit

☐ Central pain (e.g., stroke)

☐ Pathological pain (e.g., amyotrophic lateral sclerosis)

☐ Nociceptive pain (noxious stimulus from e.g., chemical, heat, mechanical sources)

☐ Mental-emotional pain

Sample conditions that may benefit from a chemotype I include

☐ HIV/AIDS-based neuropathies

☐ Severe physical withdrawal symptoms (e.g., heroin, opioids, methamphetamines)

☐ Chemotherapy-induced nausea and vomiting

☐ Spinal cord injury–induced chronic muscle spasms

☐ Alzheimer's disease with nighttime agitation or disturbed behavior

☐ Autism with agitation with self/other-harming behavior

Sample Effects with Relevance to Chemotype II

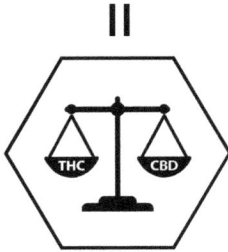

Mental/Emotional Effects

Cognition

☐ Lower chance of feeling "high" but still some potential for changes in cognition (THC-dose-dependent)

Memory

☐ Similar changes to various types of memory functions as when using a chemotype I but much milder in nature

☐ Increased CBD content reduces THC-induced memory impairments

Mood

☐ Moderate relaxation and stress reduction

☐ Low to moderate sedative effects (therapeutic in moderately overexcited/stressed nervous system-based conditions)

☐ Mood improvement (wider line between anxiolytic and anxiogenic effects)

☐ Moderate effects as anxiolytic or anxiogenic

Physiological Effects

Pain types that may benefit:

☐ Peripheral neuropathies

☐ Pain due to muscle spasms (e.g., multiple sclerosis)

☐ Cancer treatment–induced or cancer-based neuropathies

☐ Mental-emotional pain

Sample conditions that may benefit from a chemotype II include

☐ Multiple sclerosis (calms and reduces pain, spasms, urinary incontinence)

☐ Fibromyalgia (provides systemic analgesia)

Sample Effects with Relevance to Chemotype III

Mental/ Emotional Effects

Cognition

☐ No risk of getting "high," and no changes in cognition (unless your strain contains more than 0.3% THC)

Memory

☐ Potential for neuroprotection in cases of dementia improves memory function

Mood

☐ Mild relaxation and stress reduction; gentle uplift in affect (emotions)

☐ No sedative effects

☐ Mood improvements (anxiolytic with no anxiogenic effect if THC is less than 0.3%)

Physiological Effects

Pain types that may benefit

☐ General analgesic properties—possibly synergistic with opioids

☐ Inflammatory pain

☐ Chronic pain

☐ Mental-emotional pain

Sample conditions that may benefit from a chemotype III include

☐ Acne

☐ Heart disease

☐ Colitis

☐ Mental or mood disorders

    ☐ Depression

    ☐ Psychosis

    ☐ Addiction

    ☐ Anxiety

        ☐ Post-traumatic stress disorder (PTSD)

        ☐ OCD

☐ Seizure disorders

    ☐ Epilepsy (pediatric and adult)

    ☐ Dravet syndrome

    ☐ Lennox-Gastaut syndrome

    ☐ Tuberous Sclerosis complex

    ☐ Infantile spasms

If by chance you find a mismatch (in checkmark clusters) between your matches in priorities (about what kind of cannabis experience you wish to have) and the known effects listings, honor your priorities first and move on to step 2 with them in mind. You can always adjust your ingredients to include more THC later if your results do not meet with your satisfaction.

Once you have determined your basic chemotype, make a note of it and move to step 2.

Step 2: Narrowing Down to an Ideal Ratio

Step 2 addresses specific ratios of THC to CBD within each of the three chemotypes. For instance, the ACDC strain has a 1:20 THC:CBD ratio, but Harlequin is just 1:2—both are chemotype III, more CBD than THC, but at significantly different ratios. Consider that 100g of ACDC flower contains 20 times more CBD than THC, while a Harlequin flower contains only twice as much CBD than THC. You may want to think of it as "it's 20 players against 1, versus 2 players against 1." Clearly, very different odds.

In terms of eliciting therapeutic or recreational effects, the odds (or ratios) determine which effects are pronounced and which are diminished. (Common ratio samples in chemotype I may include 100:1, 50:1, 20:1, 5:1; in chemotype III may include 1:25, 1:15, 1:8, 1:4; and in chemotype II they are of course approaching 1:1.)

Many studies have demonstrated the relevance of the ratio between THC and CBD and their therapeutic implications. Newer studies reveal that longer-term co-administration has even more complex and synergistic molecular actions of special importance for patients suffering from chronic long-term diseases.[1] Given this fact, the specific ratio of THC and CBD has significant practical implications for numerous types of patients, and especially for those with stubborn symptoms that just will not go away.

Select the relevant boxes in the graphics below to see which ratio lights up for you.

Chemotype I

☐ **100:1**

Highest potential for type I effects
Very high risk of adverse effects for beginners (e.g., psychoactivity, changes in cognition)
Very low synergy with CBD

☐ **50:1**

High potential for type I effects
High risk of adverse effects for beginners (e.g., psychoactivity, changes in cognition)
Low synergy with CBD

☐ **20:1**

Moderate to high potential for type I effects
Moderate risk of adverse effects for beginners (e.g., psychoactivity, changes in cognition)
Low to medium synergy with CBD

☐ **5:1**

Moderate potential for type I effects

Moderate risk of adverse effects for beginners (e.g., psychoactivity, changes in cognition)

Medium synergy with CBD

Chemotype I (Sample) Effects Review

Mental/Emotional Effects—**Cognition:** "Feeling high," ranging from a pleasant euphoria to an unpleasant anxiety/paranoia/panic attack. **Memory:** Increases the ability to relax and be in the presence of intensely fearful memories, a positive effect in treating certain anxieties or phobias. Slows habitual memory, helpful in cases of OCD and addiction triggers, for example, can impair short-term and long-term memories (normally THC-dose-dependent and temporary). **Mood:** Deep relaxation and stress reduction. Conscious sedation (especially therapeutic in overly excited/stressed nervous system-based conditions; see sample conditions below). Significant mood improvement is possible, but there is a very narrow line between anxiolytic and anxiogenic effects, ranging from euphoria (joy) to dysphoria (feeling anxious, scared, or panicky). *Physiological Effects:* Bronchodilation (e.g., allergic asthma, cough), appetite stimulation (weight loss, anorexia, cachexia), immune-system modulation (down-regulation may be relevant to conditions in which the body's immune system begins to attack itself, e.g., rheumatoid arthritis). **Pain types that may benefit:** Central pain (e.g., stroke), pathological pain (e.g., amyotrophic lateral sclerosis), nociceptive pain (noxious stimulus from chemical, heat, mechanical sources), and mental-emotional pain. **Sample conditions that may benefit from a chemotype I include** HIV/AIDS-based neuropathies, severe physical withdrawal symptoms (e.g., heroin, opioids, methamphetamines), chemotherapy-induced nausea and vomiting, spinal cord injury-induced chronic muscle spasms, Alzheimer's disease with nighttime agitation or disturbed behavior, autism with agitation with self/other-harming behavior.

Chemotype II

☐ **1:1**

Highest potential for type II effects (refer to your type II effect checklist earlier in this chapter)

Lower risk of adverse effects for beginners (e.g., psychoactivity, changes in cognition)

Potent synergy with CBD

Chemotype II (Sample) Effects Review

Mental/Emotional Effects—**Cognition:** Lower chance of feeling "high" but still some potential for changes in cognition (THC-dose-dependent). **Memory:** Similar changes to various types of memory functions as when using a chemotype I but much milder in nature. Increased CBD content reduces THC-induced memory impairments. **Mood:** Moderate relaxation and stress reduction. Low to moderate sedative affects (therapeutic in moderately overexcited/stressed nervous system-based conditions). Mood improvement (wider line between anxiolytic and anxiogenic effects). Moderate effects as anxiolytic or anxiogenic. *Physiological Effects*—**Pain types that may benefit:** Peripheral neuropathies, pain due to muscle spasms (e.g., multiple sclerosis), cancer treatment-induced or cancer-based neuropathies, and mental-emotional pain. **Sample conditions that may benefit from a chemotype II include the following:** Multiple sclerosis (calms and reduces pain, spasm, urinary incontinence) or fibromyalgia (provides systemic analgesia).

Chemotype III

☐ **1:25**

Highest potential for type III effects

Very low risk of adverse effects (if THC content is less than 0.3%)

Very low synergy with THC

☐ **1:15**

High potential for type III effects

Slight risk of adverse effects (no risk if THC content is less than 0.3%)

Low synergy with THC

☐ **1:8**

Moderate to high potential for type III effects

Low risk of adverse effects

Low to medium synergy with THC

☐ **1:4**

Moderate potential for type III effects

Low to medium risk of adverse effects

Low to medium synergy with THC

Chemotype III (Sample) Effects Review

Mental/Emotional Effects—**Cognition:** No risk of getting "high," and no changes in cognition (unless your strain contains more than 0.3% THC). **Memory:** Positive neuroprotection in cases of dementia improves memory function. **Mood:** Mild relaxation and stress reduction; gentle uplift in affect (emotions). No sedative effects. Mood improvements (anxiolytic with no anxiogenic effect if THC is less than 0.3%). *Physiological Effects*—**Pain types that may benefit:** General analgesic properties—possibly synergistic with opioids, inflammatory pain, chronic pain, mental-emotional pain. **Sample conditions that may benefit from a chemotype III include the following:** Acne, heart disease, colitis. Mental or mood disorders: depression, psychosis, addiction, anxiety (e.g., post-traumatic stress disorder, OCD). Seizure disorders: Epilepsy (pediatric and adult), Dravet syndrome, Lennox-Gastaut syndrome, Tuberous Sclerosis complex, Infantile spasms

The reader is reminded that these ratios present naturally in cannabis flowers and are commonly utilized in full-spectrum tinctures, oils, and cannabis-containing products. Ratios can be designed by adding isolated cannabinoids to a liquid medium such as alcohol, oil, or various other forms of edibles. Most states that allow for medical cannabis require clear labeling of THC and CBD content and ratio.

Step 3: Choose the Optimal Form of Your Medicine (Inhalation, Ingestion, Topical, and Raw)

There are numerous forms, both medicinal and recreational, that contain cannabis or cannabis constituents such as smoke, vapor, ingestible, sublingual, nasal, ocular, injection, vaginal, rectal, and more (this book's appendix details a number of them). Four forms that differ in the effects they tend to produce, and the means by which they achieve them include inhalation (e.g., dried cannabis flower, cigarettes, vape pens), ingestible (e.g., extracts, tinctures, edibles), topicals (e.g., patches, creams, oils), and raw forms of cannabis (i.e., juice, leaves, flowers).

Inhalation
- Quick systemic relief
- Shorter duration
- Less dosage required
- Precise dosage requires calculations
- Titration rapid
- Minor dose-dependent risk of "high" (changes in cognition)

Ingestion
- Slow onset systemic relief
- Longer duration
- More dosage required
- Precise dosage is usually indicated
- Titration slower
- Significant dose-dependent risk of "high" (changes in cognition)

Topical
- Local tissue delivery
- Reduced systemic effects
- Longer duration
- Precise dosage is usually indicated
- Minor dose-dependent risk of "high" (changes in cognition)

Raw
- Slow onset systemic delivery
- Medium duration
- Considered food & as such precise dosing not usually indicated
- No risk of "high" (changes in cognition)

Chapter 5, Graphic 2: Major Forms of Cannabis or Cannabis-Containing Products

Inhalations of Cannabis Flower

Many patients utilize dried buds, flowers, and leaves of the cannabis plant by burning them and inhaling the resulting smoke. Some like to roll a joint and smoke it just like you would a cigarette. The upside here is ease of use; all you need is a joint and a lighter. The downside is inhalation of potentially toxic burn products and no control over the temperature setting. Some users concerned about inhaling burned decomposition products (pyrolysis), or people who want more temperature control, often opt for a vaporizer.

The reader is reminded that in many cases the dried cannabis flower or bud in front of you is still in its relatively raw composition. Therefore, depending on exposure to time, light, and heat, its cannabinoids may still primarily exist in their acid forms: THC as THCA, and CBD as CBDA. In this state the plant is not psychoactive and is limited in the physiological effects it can produce. To activate cannabis, the carboxyl group needs to be detached. When we introduce heat, for instance, decarboxylation occurs. The prefix *de-* simply means to separate from something, and in this case it is the removal of the carboxyl group that releases cannabinoids from their acid states.

The heat produced when lighting cannabis flower in the form of a cannabis cigarette (joint), water pipe (bong), or vaporizer produces decarboxylation. Inhaled cannabis spreads rapidly to the blood, brain, central tissues, peripheral tissues, and fat cells with descending speeds.[2]

While effects can come on quickly, smoking allows you to stop when you achieve your desired effect (a process called titration in medicine) or to stop if any adverse effects occur.

For instance, inhalation is great for insomnia patients who want a little help taking a nap in the afternoon, for those who want to take a temporary break from the stress of the daily routine, or those who want to take off the edge of chronic pain without affecting their cognitive functions. Inhalation is also a favorite form of micro-dosing (more on that in the next step) where people inhale small amounts to increase clarity, focus, and productivity.

For more information on the various means of inhalation, notes on vaporizing, and metered vaping, see the Appendix.

Did you know? A shallow inhalation only reaches the throat and trachea (windpipe). More specifically, an inhalation down to the two main bronchi (located between the sternal angle and the fourth thoracic vertebra) does not absorb any air. Therefore, no air or cannabis constituents can enter your system. This is not an efficient use of medicine and the resources it took to obtain it. Actual air (gas) absorption takes place in the alveoli, which is the smallest part of the lung. The human lung contains a virtual forest of alveoli (about 500 million).[3]

Ingestion of Cannabis-Infused Tinctures, Oils, and Edible Products Containing Them

An increasingly popular form of medicinal and recreational cannabis consumption is eating cannabis-containing products. Cannabis can be cooked into foods such as cookies, brownies, or even savory dishes. Cannabis is usually added to recipes in the form of an herbed butter, infused oil, or tincture. End products may include chocolate, cookies, gummi bears, cannabis-coated popcorn, and lollipops.

In contrast to inhalation, digestion of THC takes much longer to cause expected effects. However, they tend to be stronger, more intense, and longer lasting. Why? When THC passes through the liver, a portion of it is oxidized into an additional, also psychotropic metabolite (11-hydroxy-THC), which has been shown to penetrate the brain-body barrier much faster than an ingested THC molecule by itself.[4] So it takes longer (up to 1–3 hours) for ingested THC to enter and peak in blood and brain, but once converted, both THC and 11-hydroxy-THC produce a synergy and thus rapid absorption—leading to sudden and often more intense changes in body and mind (e.g., Maureen Dowd's experience). And while researchers posit that 11-hydroxy-THC may have therapeutic properties in and of itself, very little evidence about it is currently available.[5]

You should complete step 4 (dosage of THC and CBD) and step 6 (a session log) before ingesting cannabis-containing products, to avoid unintended consequences or adverse effects. Wait at least one hour (on a full stomach, two hours) before increasing the dose by the same amount. Repeat this process until you achieve the desired effect. Do not be impatient.

If you are happy with the results, your session log (step 6) will remind you of the amount it took to achieve the effect. The next time you need to use this medicine, repeat the total amount minus about 20% (to allow for changes due to time and rate of digestion) as your starting dose. For instance, ingestion tends to also work great for insomnia patients who are looking for longer-lasting effects (beyond that of taking a nap), for those with otherwise are in constant pain and who want to sleep a full night through, or for Alzheimer's patients with nighttime agitation.

Warning: Many patients, including myself, have made the classic mistake while eating cannabis-containing products and said to themselves, "I'm not feeling anything," so they keep eating until it is too late and adverse effects occur. In the case of ingesting too much cannabis, the effects may last many hours and can be intense and very unpleasant. So start slow and be patient.

Topical products: Topical cannabis products are used to deliver medicine to specific and isolated problem areas that benefit from the anti-inflammatory and analgesic properties of cannabis constituents. Concentrated cannabis oils or infused cannabis tinctures (alcohol, glycerin, or oil based) form the basis of various medicinal patches, creams, balms, and lotions.

Research has shown that the cells that make up most of the outermost layer of the skin (keratinocytes) contain a significant amount of cannabinoid receptors sites (CB1 and CB2). To continue with the lock and key metaphor introduced earlier, CB1 and CB2 function as a lock and cannabinoids (e.g., CBD, THC, anandamide, beta-caryophyllene) fit the locks with precision.

Once unlocked, the cell can realize a host of critical functions such as homeostasis (balance and stability), protection from invading microbes, and the healing of inflammations (due to allergies, injury, sunburn, dehydration, toxins, bacteria, fungi, or viruses).

But beyond these basics, relatively little hard data are known about the topical applications of cannabinoids.

Researchers from the University of Bonn, Germany[6] discovered that topical THC applied to mice significantly reduced inflammation of the skin. In 2015 scientists from the Brody School of Medicine in North Carolina demonstrated that anandamide may be an ideal, topical agent to eliminate non-melanoma skin cancers.[7]

However, it was researchers from the University of Kentucky, Lexington,[8] who provided us with the only known specific absorption rates to date. A transdermal patch containing delta-8-THC delivered a sustained and steady amount of THC to the bloodstream. The patch delivered 4.4 nanogram (ng/mL) of THC per milliliter of blood within 1.4 hours.

To put these numbers into perspective, here are estimates from the Addiction Research Center at the National Institute on Drug Abuse in Baltimore, which suggested that to "feel high," one would need to have between 7 and 29 ng/mL of THC in the blood. According to these numbers, even a topical application such as a sustained patch produces absorption below the level of "feeling high."[9]

The same study from Lexington also reported that the non-psychoactive cannabinoids CBN and CBD were able to penetrate the skin at rates 10 times that of THC, thus delivering potent anti-inflammatory and immune-modulating action to the skin without affecting the mind. Based on this evidence, topical oil application may provide a potent therapeutic synergy relevant to localized inflammation and pain management.

Thus research to date would seem to alleviate both concerns around lack of effectiveness and fears of psychoactive overstimulation from topical use of cannabinoids while encouraging the adoption of topical cannabinoid oils.

Furthermore, research has shown that the therapeutic powers of cannabis topicals are strongest when some THC (or better yet a full spectrum of plant constituents) is present.[10]

There is a key element not addressed by these transdermal study results that may also affect topical applications. When infused cannabis oil is heated and applied to the skin, absorption rates increase significantly, including that of THC making fear of "becoming stoned" a potential reality.

While no studies have been conducted to provide specific data, using THC-abundant oils at warm temperatures warrants caution:

- If by yourself, consider applying heated cannabis-infused oil only to isolated problem areas to minimize the risk of absorbing too much THC.
- If you are applying the heated oil to another person, you might want to consider wearing gloves when applying warm or hot oils to a client's skin, especially if your sensitivity is high.
- Plan ahead by creating a plan B where the client arranges for a designated driver and caretaker ahead of the cannabis oil massage appointment.
- Additionally, if the client exhibits unplanned extraordinary states of consciousness, you can do the same things that most emergency rooms do. Provide a quiet environment, reassure the client, give her company, and initiate plan B.

If you are making your own topical products or you want to create a potential synergy to amplify the expected topical effects, you may want to consider adding one or more of a number of terpenes that have been shown to significantly increase skin permeability and the movement of lipophilic cannabinoids (lipophilicity) across the skin membrane such as nerolidol, limonene, linalool, geraniol, or terpinol.[11]

Raw cannabis

Many grower-patients consume fresh cannabis leaf as a salad mixed with other greens or a component of a smoothie. Fresh raw leaves can also be juiced; this is often diluted with vegetable juices to disguise its bitter taste. I personally like to add apple juice.

Fresh cannabis contains CBD and THC in their acid forms. THCA (tetrahydrocannabinolic acid) is not psychoactive in fresh, raw cannabis, so this may be an ideal option for patients who need or want to avoid mind-altering effects while engaging the raw plant's potentially therapeutic properties. Research in this area is at an early stage, but results indicate a possible role for THCA in cases of nausea and vomiting,[12] neuroprotection,[13] and anti-inflammatory effects.[14]

CBDA (cannabidiolic acid) has similarly demonstrated its ability to mitigate nausea and vomiting.[15–17] It has also been shown to offer a therapeutic avenue in reducing breast cancer cell migration, including that of aggressive breast cancer cell lines.[18]

Step 4: Determining Your Proper Dose/Amount

If you have never used cannabis before and wish to prevent adverse effects, or if you are looking to realize and sustain a specific therapeutic or recreational effect, you must know the actual amount of THC. Step 4 helps you determine the precise amounts of THC as well as that of CBD.

There is one caveat. The process to determine actual THC content varies between inhaled flower and ingestible forms of cannabis. Your flower is calculated based on THC percent by weight, while a cannabis-containing tincture or oil is measured as milligrams of THC per milliliter of tincture or oil.

For most ingestible forms sold in dispensaries, the THC amount is given as a measure such as milligrams per dropper or the amount of THC in the whole container. Most producers of ingestible items also provide THC to CBD ratios on the label. In fact, most states require consistent testing and labeling of both key cannabis constituents (THC and CBD). Therefore, an ingestible form is easier to use than cannabis flower, which requires a couple of calculations to determine precise THC and CBD amounts.

Step 4a: Ingesting

While the THC amount in ingestibles ought to be listed on the label (as many states mandate), the effects produced are more complex to titrate, that is, to measure the precise dose responsible for the desired effect. However, once you know your optimal dose from trials, the advantage to ingesting edibles is that effects tend to have a longer duration, and potential damage from smoking is avoided. Of course, this assumes that the products you eat are consistently manufactured.

Taking into account the variables of absorption (typical loss rates during digestion; estimation of impact of 11-OH-THC), research suggests an ingestion:inhalation ratio of about 1:6.[19] That is, for every 1 mg of ingested THC, about 6 mg of inhaled THC may be necessary to achieve a similar effect. The reader is asked to keep in mind that this ratio is an estimate, to be considered with appropriate caution. It was derived mathematically from averages of different study results, each with ranges that had large margins.

Cannabis Ingestion Dosing Considerations

As always, start slow and small. Consider starting with 2.5 to 5 mg THC increments.

> Numerous publications use the abbreviation ml for milliliter. However, this book uses the capital L in mL to avoid confusion with the roman numeral I.
>
> 1 g = 1,000 mg
>
> 1 L = 1,000 mL
>
> **1 drop.** You can measure volume by drops. In medicine intravenous (IV) droppers are measured by the amount of mL delivered per single drop. A macro drip IV rate delivers 10 drops per mL. A micro drip delivers 60 drops per mL. Similarly, a dropper bottle delivers volume depending on the size of the dropper. A 1-oz. bottle dropper contains about 0.05mL per drop; and about 20 drops make 1 mL.
>
> **1 dropper-full.** If you want to know how many milliliter are in a dropper-full, fill it up as you naturally would by squeezing the rubber top (you can never get it to fill up completely) and count the drops as you slowly empty them back into the bottle. Whatever your number is, multiply it by 0.05 mL and you have an estimated "dropper-full" milliliter amount.

Example 1: You are at the dispensary looking at your tincture options. What to do?

Most dispensaries offer tested products that list measurements of liquids in mg/mL or total number of mg per bottle. Using 300 mg (THC)/30 mL (1 oz.) as an example.

If your preferred dose is 10 mg (THC), then you would take 1 mL per dose (30 doses per bottle); 300 mg (THC)/30 mL = 10 mg (THC)/mL.

If, however, you need 40 mg (THC), then you would take 4 mL (7.5 doses per bottle).

If your dose is less than the amount in 1 mL (10 mg THC in this case), you can use drops to fine-tune your specific dose. For example, if you need 5 mg you can measure 10 drops (=1/2 mL THC).

10 mg/mL —> 1 mL/20 drops = 0.5 mg/1 drop.

Example 2: You are making your own tincture at home. How to measure with precision?

People interested in making their own infused tincture or infused oil can use the following example. You can use the infusion straight or measured accurately into homemade edibles (whatever the form).

For instance: If a tested quantity of 1 g (1,000 mg) cannabis contains 20% THC, then the total sample contains ~200 mg of THC. If we want to mix 20 g of this particular flower with 100 mL of coconut oil, for instance, we must convert the grams into milligrams, which is 20 × 1,000 = 20,000 mg.

1 g (1,000 mg) —> 20% = 200 mg

20 g (20,000 mg) —> 20% = 4 g (4,000 mg)

4,000 mg THC ÷ 100 mL (coconut oil) = 40 mg/mL

In this example, each 1 mL will deliver 40 mg of THC (1 mL = 40 mg of THC).

If 40 mg is sufficient to induce desired effects in terms of symptom reduction or mood improvements, you will need to measure out 1 mL for every dose of 40 mg of THC.

If the dropper delivers 1 mL as 20 drops, then each drop delivers 2 mg of THC.

40 mg/mL —> 1 mL/20 drops = 2 mg/1 drop.

If you only need 10 mg THC per dose and each drop contains 2 mg, then you would need 5 drops to get 10 mg of THC.

Ingestible Cannabis Dosing Considerations Summary

- As always, start slow and small.
- Consider starting with no more than 5 mg increments of THC. How?
- Figure out how many drops equal 5 mg (THC) and take that amount.
- If an edible (cookie, brownie, chocolate) contains 20 mg (THC), divide in 4 quarters and eat one.
- In both cases wait for one hour (two hours if stomach is full).
- Notice and note any changes in your journal.
- If no result or not enough progress, increase by same amount.
- Wait for one hour (two hours if stomach is full).
- Notice and note any changes in your journal.
- Repeat until you achieve the desired effect.
- Note in your journal exactly how many drops you took.
- You may consider this total amount of drops, minus about a quarter (to account for time passed in previous experiment), as a possible starting dose the next time around.
- Keep track of any changes in your work pages and adjust accordingly.
- Stop and regroup if you notice unwanted effects.

Step 4b: Inhalation

For inhaled cannabis flower, we need to consider percent by weight concentration of THC and CBD. Because of epigenetics—that is, environmental changes that switch some genes on and others off without affecting actual DNA sequence—concentrations can differ significantly from one plant to the next, even though they both started with the same clones or seeds.

In the health sciences, concentration defines and measures the action of a cannabinoid—for instance, how strong a concentration we need to produce an effect of a given intensity. Concentration is often measured in percentage by weight.

How much percentage and milligram by weight of THC (or CBD) does my cannabis flower contain, and how much of it will I actually absorb to produce the therapeutic effect I need or want?

The answer to this question is not straightforward and requires a two-step calculation (one to figure out actual percentage and second the actual amount of milligrams by weight of THC, CBD or any other cannabinoid. While there are a couple of methods you can use to approximate an accurate result, the answer is never 100% precise because of variables beyond our control such as loss of vapor or smoke when inhaling or the number of cannabinoid receptor sites expressed in different patients even if they have the same diagnosis. However, with a few calculations we can come pretty close and produce precise and sustainable effects.

The reader is reminded that in this book the THC and CBD values used throughout are based on already decarboxylated or activated cannabis flowers. Figuring out how much THC or CBD is in your cannabis flower before decarboxylation occurs requires a closer look and a bit of additional calculating. Since there is more than one formula and no current industry-wide standard, results may vary somewhat, depending on testing facility or producer preferences. Here are some of the relevant items (and their values) commonly listed (depending on producer/dispensary preference or state policy), followed by an example.

Sample Label

- **Strain name:** OG Kush
- **Net weight:** Weight of your cannabis flower without the package materials (e.g., 3 gm)
- **Lot/Batch number:** 07031958 (to track product throughout the distribution chain)
- **Date of harvest:** March 2019
- **Tested by:** Georgia Laboratories LLC
- **Produced by:** Tasmanian Devils Inc.
- **Information about method of cultivation:** outdoor, indoor, hydroponic, etc.
- **The potency analysis:** May include some or all of the ingredients and percentage values, as listed below:
  - Total cannabinoid content 17%
  - Total activated cannabinoid content 1%
  - THC 1%
  - THCA 20%
  - THC 1% decarboxylated (activated)

Remember, at this stage (when you buy your flower in the dispensary), the cannabinoids such as THC in your flower is mostly in its acid form (THCA) and still must be activated by heat to become psychoactive. Therefore, the actual activated (decarboxylated by time and drying) level of THC is usually a very low number (in this sample 1%).

Most people assume that you would add 1% THC to 20% THCA to arrive at your total percentage of THC. However, the actual amount will be less. This may sound counterintuitive, but here is why: THC is lighter than THCA because, one, it loses its carboxyl group during the heating (decarboxylation) process; and two, some of the THC has converted into CBN. So the actual amount of total THC (17% in this example) will be less. How much less will also depend on the formula used to calculate it. Here are two of the more common formulas that show the differences of actual percentage calculations:

Method/Formula 1 is simple but also the least accurate: $\boxed{\text{THCA\% + THC\%}}$

Example 1: 20% + 1% = 21% Total THC

Method/Formula 2 is a bit more math but more accurate: $\boxed{(0.8 \times \text{THCA\%}) + \text{THC\%}}$

Example 2: (0.8 × 20%) = 16% + 1% = 17% Total THC

As you can see, the difference in accuracy (in this case 4%) can make a difference in the therapeutic or recreational effects you want to realize.

The difference in formulas is not the only variable that affects how much of the cannabinoids in your flower actually make it into your system. Others include the temperature used to heat your cannabis and the type of heating device used (e.g., cigarette, bong, vaporizer). Keep in mind also that with each temperature change come variations in the effectiveness of the plant's other constituents, which tend to have optimal heating ranges that can differ from that of THC. Further, other variables that cannot be controlled include actual absorption rates, speed of crossing the blood-brain barrier, differences in endocannabinoid deficiencies, or differences in individual endocannabinoid tone for instance.

Remember, to know the actual amount of THC by weight you still need to convert your total THC percentage into milligrams by using this formula (outlined in more detail in chapter 5, Step 4):

$$\frac{\text{\# flower in mg } \times \text{ \# THC\%}}{100\%} = \text{total \#THC mg}$$

Now plug in the abovementioned % values to determine mg THC by using a sample amount commonly used in a vaporizer bowl: 50 mg of cannabis flower.

Method/Example One: 50 mg × 21%THC ÷ 100% = 10.5 mg THC

Method/Example Two: 50 mg × 17%THC ÷ 100% = 8.5mg THC

Comparing the two formulas above, the actual difference in THC by weight is 2 mg, which with some patients is more than enough to produce an adverse effect.

The same idea applies to other cannabinoids such as CBD, CBN, or CBG.

It is easier with ingestibles such as oils, cookies, and so on where the total amount of THC will be given already by weight: this cookie contains 10 mg THC, or one dropper-full contains 3 mg THC.

Measuring Decarboxylation

Cannabis produces the primary cannabinoids in their acid or carboxyl forms such as Δ9-tetrahydrocannabinolic acid (THCA), cannabidiolic acid (CBDA), and cannabigerolic acid (CBGA). While the acid forms have unique and valuable properties in their own right, such as the ability to inhibit the growth of highly aggressive breast cancers,[20] many of the vast physiological responses of the human body to these cannabinoids are only realized when the substances are converted or transformed by a process called decarboxylation. This process is initiated through a chemical change induced by heat, and more slowly by light, or oxygen. When exposed to one of these, the carboxyl group detaches from the cannabinoid acid forms and thereby becomes decarboxylated and available for chemical activity in the human body. You may want to think of it as an activation.

Past methods of analyzing cannabis decarboxylation were less efficient and less accurate than today's methods of using high-performance liquid chromatography (HPLC). A study conducted by the University of Mississippi in late 2016 using the latest technology—an ultra-high-performance supercritical fluid chromatography (UHPSFC)—provides the currently most sensitive analysis of the cannabinoid decarboxylation processes and the temperatures at which they occur most efficiently.[21]

To access THC or CBD and release their therapeutic potential, both THCA and CBDA must be fully decarboxylated. Here is the latest data using UHPSFC:

- At less than 100°C (212°F) decarboxylation was still incomplete under 60 minutes.
- At 110°C (230°F) decarboxylation was completed in 30 minutes.
- At 130°C (266°F) decarboxylation was completed in 9 minutes.
- At 145°C (293°F) decarboxylation was completed in 6 minutes.

Based on current market trends, the average concentration of THC in dried cannabis flowers can vary significantly.

Here is a practical example to explore concentrations of 50 mg of Purple Urkle cannabis flower. A chemotype I such as a Purple Urkle with a THC:CBD ratio of 20:1 at 7% THC concentration contains 3.5 mg THC and about 0.2 mg CBD. In contrast, a Purple Urkle with the same 20:1 ratio at 30% THC contains 15 mg THC and 0.8 mg CBD. Both are categorized as chemotype I and have the same 20:1 ratio of THC to CBD, but they deliver significantly different concentrations, which directly influences any resulting effects.

Remember that cannabinoid absorption from inhalation to crossing the blood-brain barrier is nearly instantaneous, whatever the concentration, so be patient. For some people, the line between an effective dose that really meets their needs and an adverse effect can indeed be very thin. Continue this slow, careful, observant process until you have achieved your stated goal of symptom reduction or mood improvement. Do not inhale more. More is not better.

Always stop if you are experiencing unwanted effects. In the case of very mild adverse effects, wait a full day and begin again. In the case of moderate or severe adverse effects, your options include switching to a lower concentration of THC, a lower sub-ratio of THC, or to another chemotype with lower THC altogether.

If you were happy with the results—for example, it took two medium-size inhalations to feel the desired therapeutic effect—use two same-size inhalations for symptom relief in the future. This is your current subjective therapeutic window or optimal dose with this delivery mechanism and type of cannabis. Note it in your journal.

Which Method of Vaporizing is Recommended?

For patients, especially those dealing with chronic conditions, the use of a vaporizer offers near-instant bioavailability of cannabis constituents, making it possible to control and measure your progress in terms of symptom reduction and mood improvements within minutes (as with smoking). Thus you reduce the risk of "impatience-induced" adverse effects (i.e., eating more while waiting for the effects to occur). Vaporizing also has the additional benefit of significantly reducing the risk of inhaling smoke (combustion products). Considering a vaporizer to dispense your medicine is a strong option. Numerous products offer a variety of means to achieve vaporization. Costs of a vaporizer can range from less than $100 to more than $1,000.

The ideal temperature setting when using a vaporizer is being debated by different schools of thought. It roughly ranges from ~170°C to 230°C (~338°F to 446°F) with a midpoint between them being 200°C (392°F). Some advocate for a lower temperature setting (for a lighter, more functional experience with a focus on terpenes), while others make the case for a higher temperature that releases more cannabinoids (for a more intense, heavy relaxation and psychedelic experience).

A recent study from Switzerland tested a variety of vaporizers, and results point to the following:[22]

1. The temperature used for this experiment was 210°C (410°F).
2. The tested electrically driven vaporizer produced efficient decarboxylation rates and reliably released cannabinoids into the vapor.
3. An electrically driven vaporizer (instead of a gas-powered one) obtained the cleanest vapor.
4. Those devices that produced a cooler vapor produced a milder (less irritating to the throat) inhalation experience but were less efficient in producing the maximum amounts of cannabinoids released. In contrast, devices that were more efficient in turning cannabinoids into vapors produced a harsher inhalation experience. The takeaway? Your choice: slightly less cannabinoids but easy on the throat, or maximum efficient use of cannabinoids with a harsher inhalation. Something to consider if cost is an issue, or if you have a throat condition that appreciates a very gentle approach.

The tested electrically driven vaporizers included:

- Volcano Medic®—Total THC and CBD released into vapors: 58.4 and 51.4%
- Plenty Vaporizer®—Total THC and CBD released into vapors: 66.8 and 56.1%
- Arizer Solo®—Total THC and CBD released into vapors: 82.7 and 70.0%
- DaVinci Vaporizer—Total THC and CBD released into vapors: 54.6 and 56.7%

If you are using a vaporizer, you may notice that the initial inhalations contain more flavors and scents, while the latter inhalations seem almost bland by comparison. Terpenes, the substances that give plants their odor, tend to evaporate more rapidly and at lower temperature; thus you are likely to notice the flavors and scents more in the beginning.

In contrast, cannabinoids are silent in their flavor and scent, and once the terpenes are gone, taste and scent tend to become flat. However, this does not mean that cannabinoids are also gone. To realize more efficient utilization of cannabinoids, consider inhaling past the point where you notice a significant drop of flavor or scent. Keep track of your progress in your journal, and soon a pattern will emerge that informs you about when inhaling produces no shifts in your experience, thus indicating exhaustion of this amount or dose. You may want to remember the number of inhalations (by noting it in your journal) to better guide your future decisions.

To date, there is very little objective quantitative information available that demonstrates just how many vaporized inhalations it would take for a given amount—such as 100 mg of cannabis flower containing 14 mg of THC—to become depleted of its cannabinoids. One laboratory (MCR Labs) published a case study using a vaporizer called Magic Flight Launch Box®. The cannabis flower used was an indica-dominant hybrid called Mango. Results showed that it took about 40 10-second inhalations, at the rate of 0.3 mg THC, for the sample to be depleted of THC. While the authors did not disclose the total amount of THC contained in their sample flower used, the math suggests a starting sample of 12 mg THC (0.3 mg THC × 40 10-second inhalations = 12 mg THC).[23]

Dosing Considerations When Using Unknown and Untested Cannabis Flowers

- When inhaling flowers of unknown chemotype, ratio, or concentration:
- The only control you have is the number and size of inhalations. So start very slow and small.
- Consider starting with 50 mg of cannabis flower. How?
- Take 1 gram for example. (I am using more than the typical daily amount, to make the math more obvious.)
- Subdivide into 4 portions (4 × 250 mg).
- Further subdivide into 5 more portions (approximately 5 × 50 mg).
- Sprinkle into your cigarette or load your vaporizer with 50 mg.
- Inhale about two thirds of the way into your lungs (without straining yourself).
- Continue the last 1 third with room air (to pull vapor past the trachea).

- Do not strain or hold your breath.
- Gently exhale.
- Stop and wait two to five minutes.
- Notice and note any changes in your session log.
- Repeat until you achieve the desired effect.
- Note in your journal exactly how many inhalations you took.
- Stop and regroup if you notice unwanted effects.

To obtain precision results, to avoid adverse effects, and to sustain desired effects of symptom reduction and mood improvements, you need quality-based verifiable test information about what you are consuming.

Micro-dosing Cannabis

Through a concept popularized in part by Albert Hofmann—the chemist who invented LSD, and more recently by Silicon Valley employees following in the footsteps of Steve Jobs—a micro-dose of LSD (to support creativity, productivity, and a more vivid imagination) has become the model for a novel approach to working with cannabis and cannabis-containing products.

Some people use the term *micro-dosing cannabis* to simply mean a tiny amount, while others imply taking a little less than you normally would to feel an effect. Micro-dosing may not be useful when dealing with severe pain, disability, or cancer. But for people doing creative work, those needing a little help to focus, or those who wish to continuously support and balance their endocannabinoid tone in a very subtle way, this approach might just be for you.

By and large, scientific evidence regarding micro-dosing is only beginning to emerge. Two studies may help to develop a concept of the effects of smaller dosages of cannabis, one from the University of California, Davis Medical Center (2013),[24] and another from the University of Bonn, Germany (2017).[25]

The double-blind, placebo-controlled trial at UC Davis examined the effects of low- and medium-dose cannabis administration in the treatment of neuropathic pain. Researchers obtained cannabis grown by the government at the University of Mississippi under the supervision of the National Institute of Drug Abuse, which tends to contain THC concentrations at the lowest of the four levels identified in the CannaKeys Process, or up to ~7%. A total of 0.8 g (800 mg) of cannabis was placed in a Volcano vaporizer, vaporized, and collected in the vape bag.

Over the years, numerous inhalation methods have been advocated; for this experiment, researchers instructed the patients to inhale for 5 seconds, hold for 10 seconds, exhale, and wait for 40 seconds before repeating the inhalation procedure for a total of four inhalations. Three hours later patients repeated the procedure but had the option to inhale up to eight times. The low dose of cannabis contained ~1.3% THC, and the medium dose contained ~3.5% THC.

When 800 mg of flower contains 1.3% THC, the total amount of THC contained in that sample is 10.4 mg (1.3% × 800 mg ÷ 100% = 10.4 mg).

When 800 mg of flower contains 3.5% THC, the total amount of THC contained in that sample is 29.6 mg (3.7% × 800 mg ÷ 100% = 29.6 mg).

Keep in mind that the total amount was spread out over two sets (three hours apart) of inhalations (four or eight times). Given the loss of cannabis constituents during vaping, inhalation, actual absorption, and metabolism over time, the actual amount of THC that made it into blood plasma and across the blood-brain barrier was indeed relatively small. The takeaway from these experiment results showed that even a low dose was as effective as the medium dose at any given time during the experiment.

No information was provided about the presence of CBD or other cannabinoids or terpenoids. Therefore, chemotype and ratio information was lacking (most likely the cannabis used was a chemotype I).

The Bonn study discovered that a chronic low dose of THC was able to rejuvenate the brains of aging mice. Based on prior evidence that demonstrated significant endocannabinoid modulation in the processes of aging, scientists set out to understand more about how the ECS works to slow or reverse the aging process.

As far as mouse lifespan is concerned: 2 months is considered young, 12 months mature, and 18 months old. Using one health care marker of aging (hippocampal gene transcription patterns), researchers showed that the administration of chronic low-dose THC (3 mg/kg/d×28d) turned the aging clock back from old to young and thus returned memory and cognitive performance comparable to that of young mice. Not a bad accomplishment.

To begin micro-dosing, you may find it helpful to alert your endocannabinoid system of coming changes and intentions by pushing the "reset button." In other words, avoid the use of any cannabis-containing medicine for a "washout period" of about five days[26] to achieve a natural baseline endocannabinoid tone.

How Do I Start Micro-dosing Cannabis?

The idea here is to use an amount of cannabis or cannabis-containing products with small enough amounts of THC to avoid noticeable shifts in cognition. The exact amount is fluid and largely depends on you. For some, a small amount of 0.1 mg to 0.4 mg of THC is right, while for others a microdose means using a bit more. To provide some context, the reader is reminded that the pharmaceutical version of the oromucosal spray Sativex suggests a starting dose of 2.7 mg THC and 2.5 mg CBD with a single spray beneath the tongue.

Key Points, Steps 4a and 4b

Concentration of cannabinoids in the cannabis flower significantly determines the effect you will experience.

When inhaling flowers of a known and tested content: You can make informed decisions—for example, if you are very concerned about getting high but want to include THC, you may want to consider starting with a relatively low amount (2.5-5 mg) of THC.

For instance, if 50 mg of flower with 5% THC it follows that it contains 3.5 mg THC. In contrast, if 50 mg of flower contains 21% THC, then that sample contains 10.5 mg THC, so you want to choose 50 mg of the first flower containing 5%THC.

When inhaling flowers of an unknown chemotype, ratio, or concentration: Start very slow and small. Consider using 50 mg flower increments. Stop as soon as you notice any unwanted effects. Remember: To obtain precision results, to avoid adverse effects, and to sustain desired effects of symptom reduction and mood improvements, you need quality-based verifiable test information about what you are consuming.

Ingesting cannabis-infused tincture or oils (or products that contain them): Consider starting with 5 mg (or less) increments of THC. Wait for one hour (two hours if your stomach full). Notice any changes in your journal. Repeat until you achieve the desired effect. Note in your session log exactly how many drops you took. Stop and regroup if you notice unwanted effects. Remember: This conversion ratio is a very rough guideline, and each person may react differently after trying the conversion ratio of 1:6 (ingested vs. inhaled).

Micro-dosing: Use an amount small enough to avoid noticeable symptoms or changes in mood. The exact amount is fluid and largely depends on you. However, you may consider the following.

THC Dosage Considerations (based on dosages used in the currently available human trials listed in chapter 3, section THC)

THC micro dose: ~0.1 mg to 0.4 mg (0.001 mg/kg to 0.005 mg/kg)

THC low dose: ~0.5 mg to 5 mg (0.006 mg/kg to 0.06 mg/kg)

THC medium dose: ~6 mg to 20 mg (0.08 mg/kg to 0.27 mg/kg)

THC higher dose: ~21 mg to 50+ mg (0.28 mg/kg to 0.67 mg/kg)

(For some people an upper limit is 1,000+ mg)

CBD Dosage Considerations (based on dosages used in the currently available human trials listed in chapter 3, section CBD)

CBD low dose: 0.4 mg to 19 mg (0.005 mg/kg to 0.25 mg/kg)

CBD medium dose: 20 mg to 99 mg (0.26 mg/kg to 1.32 mg/kg)

CBD high dose: 100 mg to 800+ mg (1.33 mg/kg to 10.7 mg/kg)

(upper limits tested ~1,500 mg)

Formulas for calculating cannabinoid % before decarboxylation (activation)

Method/Formula 1 is simple but also the least accurate: THCA% + THC% = THC%

Method/Formula 2 is a bit more math but more accurate: (0.8 × THCA%) + THC% = Total THC%

Formula for calculating mg cannabinoid from % cannabinoid (of already decarboxylated flower): First, convert ounce (oz.) or gram (g) amount to milligram (mg) 1 oz. = ~28,350 mg, 1g = 1,000 mg. Second, choose your mg amount (samples below are based on 50 mg flower—a common loading dose). Third, plug your values into the formula:

flower in mg × # THC% = total #THC mg

Step 5: Designing the Entourage Effect

The term *entourage effect* describes a relationship among various cannabis constituents. One compound may show no activity alone, but it becomes biologically active in the presence of two other related compounds, especially in the context of a very specific patient population. Thus, over time, an entourage effect can be defined as one that enhances or magnifies therapeutic effects in body, mind, or emotion; produces balance among organ systems (homeostasis); enhances efficacy; and/or reduces adverse effects. For instance, the terpene myrcene enhances the sedative effects of THC, while CBD tames the psychotropic effect of THC. For patients dealing with obesity, metabolic syndrome, and diabetes in addition to needing effective analgesia, the ratio of THCV (to increase satiation) to THC (for analgesia) becomes of paramount importance. Another example includes the combination and optimal ratio of THC and THCA in the treatment of autism and certain inflammatory conditions.

In addition, cannabis also produces terpenes, which may be present in significant quantity or in trace amounts. Researchers posit that, based on current analytical limits and technological measuring capabilities, a threshold of **terpene concentrations of 0.05%** or higher is necessary to expect pharmacological effects in humans.[27]

Terpene-based effects tend to be less potent when compared to THC and CBD, but they do have independent therapeutic effects that can be used to amplify or reduce cannabinoid-induced effects. In the following section we focus on the synergistic effects of terpenes, major cannabinoids, and the acid forms of THC and CBD.

	# of Studies	Anti-oxidant	Anti-cancer	Anti-microbial	Anti-inflammatory	Breathe Easy	Cholesterol	Immune-modulating	Neuroprotective	Analgesic	Relaxing	Sedating	Sleeping Aid	Spasm/Seizure	Uplifting
Caryophyllene	●	✓	✓	✓	✓		✓	✓	✓					✓	✓
Limonene	●	✓	✓	✓		✓					✓				✓
Linalool	◐		✓	✓	✓					✓	✓	✓			✓
Eucalyptol	◐	✓	✓			✓				✓					✓
Phytol	◐	✓		✓	✓					✓	✓				
Nerolidol	◐		✓									✓		✓	
Myrcene	○		✓	✓						✓	✓	✓	✓		
Pinene	○		✓	✓		✓			✓						✓
Camphene	○						✓		✓						
Terpineol	○		✓								✓				
Terpinolene	○											✓	✓		
Borneol	○	✓							✓						
Geraniol	○	✓						✓	✓						
Humulene	○			✓											

Chapter 5, Graphic 3: Designing an Entourage Effect with Common Cannabis Terpenes

I have never worked with an entourage effect. Where to start?

People tend to work with terpenes in different ways. One way is to visit a dispensary and ask for cannabis flower with the desired chemotype, ratio, potency and terpene profiles. Most patient-focused dispensaries tend to have a variety of options available. Which entourage effect is on top of your list, which one second, which one third? Which of your narrower options available at this dispensary has the highest concentration of the terpene or terpene profile you are looking for? If you cannot match all the terpenes you like, go with number one on your list. You are done with step 5.

Another method is simply to purchase a small bottle of high-quality terpene (at concentration safe for human consumption or topical use). After you inhale the smoke or vapor of your specific, precise cannabis flower, inhale the aroma of that terpene from the bottle that is number one on your list of desired entourage effects. You may want to alternate inhaling the terpene scent and inhaling your vaporized cannabis flower. Also, consider using an essential oil diffuser filled with an essential oil abundant in the terpene of your choice. (For a list with essential oils rich in specific terpenes, see appendix, graphic 5).

Another method popular in the recreational sector involves dabbing, that is, vaporizing and inhaling solvent-extracted cannabis concentrates such as butane hash oils or CO_2 extracted oils. Each of these concentrates has unique production methods and comes with different looks and textures and are named accordingly: wax, shutter, budder, or crumble. However, no matter what concentrated product is used, what matters to the recreational connoisseur and discerning patient alike is that a tiny amount of the concentrate is dipped into a specific terpene or terpene profile just before being heated and inhaled, providing users with the unique flavor of the terpene or terpene profile of their choice.

For those who wish to create an entourage effect geared toward symptom reduction or mood improvements, graphic 3 in this chapter can serve as a reference guide.

On the right side of the graphic, you see some evidence-based examples of commonly observed terpene-induced effects. The circle icons indicate the rough number of select scientific studies conducted on each terpene. The darker the saturation, the more studies have been completed—thus indicating the degree of interest and degree of the substance of the evidence of associated effects.

Below is a summary highlighting these terpenes, relevant conditions, and symptoms that select research has suggested may provide additional ways to support your journey toward optimal health.

For more information, annotations, and the science behind these summaries, see the Appendix: The Science Behind the Top 10.

Monoterpenes

- **Borneol**
  Effects on the body: antibacterial (middle-ear infection), analgesia via TRPM8.
  Effects on mind and emotion: none reported.

- **Camphene**
  Effects on the body: analgesia (via TRPV3), may reduce hyperlipidemia and related conditions (e.g., cardiovascular illness, cholesterol-based gallbladder stones).
  Effects on mind and emotion: none reported.

- **Eucalyptol (aka 1,8 cineole)**
  Effects on the body: reduces phlegm, cough, and nasal congestion; produces analgesia; may be protective against certain cancers (leukemia).
  Effects on mind and emotion: better concentration, improved cognition and mood.

- **Geraniol**
  Effects on the body: antioxidant, neuroprotective, eases neuropathies, possibly inducing vasorelaxation.
  Effects on mind and emotion: none reported.

- **Limonene**
  Effects on the body: may protect against breast and prostate cancer, antioxidant, may reduce obesity and induce weight loss, antifungal, may help with cholesterol-based gallbladder stones, may normalize peristalsis in cases of heartburn and GERD.

Effects on mind and emotion: uplifting, antidepressant, anti-stress, relaxing, anxiolytic, calming effect may help with insomnia.

- **Linalool**
  Effects on the body: anticonvulsant, antiparasitical (leishmania), may reduce high blood pressure, may produce synergistic analgesic effects with opioids, may protect against certain cancers (kidney, breast, lymphoma, leukemia), and may produce synergistic effect with certain chemotherapy drugs.
  Effects on mind and emotion: relaxing, calming, anxiolytic.

- **Myrcene**
  Effects on the body: possible sleeping aid, analgesia, may prevent peptic ulcer diseases.
  Effects on mind and emotion: sedating, heavy, sleepy, couch-lock effect, relaxing.

- **Pinene**
  Effects on the body: bronchodilation, antifungal (candida), antibacterial (MRSA), may help Alzheimer's disease, Parkinson's disease, and multiple sclerosis, may inhibit liver cancer.
  Effects on mind and emotion: focus, alertness, improved memory.

- **Terpinol**
  Effects on the body: antiproliferative effects against cancer cell lines such as lung cancer and certain leukemia cells (erythroleukemic K562); enhances the permeability of skin to lipid-soluble compounds such as cannabinoids.
  Effects on mind and emotion: calming, relaxing.

- **Terpinolene**
  Effects on the body: sedative.
  Effects on mind and emotion: calming, sleepiness, sedative effects.

Sesquiterpenes

- **Beta-Caryophyllene**
  Effects on the body: immune modulation, anti-inflammatory, neuroprotective (spasms, seizures, epilepsy; induces neuritogenesis); may protect from Alzheimer's disease and vascular dementia; may be beneficial in cases of colitis; antibacterial (tuberculosis, leprosy, *Staphylococcus aureus*, *Pseudomonas aeruginosa*); antifungal (*Candida albicans*, *Aspergillus niger*); may be protective against Type II diabetes; may have anti-cancer properties (kidney, colon, pancreatic, melanoma, lymphoma); may produce a synergistic effect with certain chemotherapy agents (e.g., Paclitaxel), analgesic effects (in cases of inflammatory and neuro-pathic pain), antioxidant, may protect against binge drinking or chronic alcohol consumption–induced alcoholic steatohepatitis, nonalcoholic fatty liver disease, and steatosis (fatty liver diseases). Has been shown to prevent and treat periodontal disease in dogs.

Effects on mind and emotion: mood improvements (anxiolytic, antidepressant), no mind-altering effects.

- **Caryophyllene oxide**
  Effects on the body: possible anticancer properties (cervical, leukemia, lung, gastric, ovarian), may be synergistic with chemotherapy drugs; antifungal (onychomycosis).
  Effects on mind and emotion: none reported.

- **Humulene**
  Effects on the body: anti-inflammatory.
  Effects on mind and emotion: none reported.

- **Nerolidol**
  Effects on the body: sedative, antispasmodic, antifungal (*Microsporum gypseum*), antiparasitical (babesiosis, malaria), may produce a synergy with pharmaceutical antibiotics.
  Effects on mind and emotion: possibly mildly sedating.

Diterpenes
- **Phytol**
  Effects on the body: possibly analgesic, antioxidant, anti-inflammatory, antimicrobial (*Staphylococcus aureus*), antiparasitical (schistosomiasis).
  Effects on mind and emotion: relaxing effects.

Major Cannabinoid Groups (in alphabetical order)
- **CBC**
  Effects on the body and synergistic terpenes:
  Antibacterial—pinene, phytol, beta-caryophyllene
  Effects on mind and emotion:
  Antidepressant—eucalyptol (aka 1,8 cineole), limonene, beta-caryophyllene

- **CBD**
  Effects on the body and synergistic terpenes:
  Alzheimer's disease—pinene
  Parkinson's disease—pinene
  Synergistic analgesia with opioids—linalool
  Antibacterial activity against MRSA—pinene, phytol, beta-caryophyllene
  Lung cancer—alpha-terpineol
  Breast cancer—limonene, beta-caryophyllene oxide
  Analgesic effects in neuropathies—geraniol
  Acne—alpha-terpineol (to enhance skin permeability), limonene, beta-caryophyllene
  Neuroprotective—geraniol, beta-caryophyllene

Antioxidant—limonene, geraniol, beta-caryophyllene, phytol

Effects on mind and emotion:

Lowers heart rate—geraniol (vasorelaxation)

Mood improvements—beta-caryophyllene, eucalyptol (aka 1,8 cineole)

Arousal—consider geraniol (vasorelaxation), limonene for relaxing and uplifting effects

Anxiolytic—limonene, linalool, beta-caryophyllene

- **CBDA**
  Reduces nausea and vomiting
  Mitigates breast cancer cell migration

- **CBG**
  Effects on the body and synergistic terpenes:
  Muscle relaxation—geraniol
  Analgesia—linalool, beta-caryophyllene, borneol, camphene, eucalyptol (aka 1,8 cineole), myrcene, phytol
  Sedation—myrcene, terpenolene, nerolidol
  Antibacterial (MRSA)—pinene, phytol, beta-caryophyllene
  Prostate cancer—limonene
  Anti-leishmanial—linalool
  Effects on mind and emotion: none reported.

- **CBN**
  Effects on the body and synergistic terpenoids:
  Antibacterial activity against MRSA—pinene, phytol, beta-caryophyllene
  Effects on mind and emotion: relaxation, sedation, and sleep—myrcene

- **THC**
  Effects on the body and synergistic terpenes:
  Analgesia (via CB1 and CB2)—beta-caryophyllene
  Antibacterial (MRSA)—phytol, beta-caryophyllene, pinene
  Immunomodulation—beta-caryophyllene
  Bronchodilation—pinene
  Effects on mind and emotion:
  Stress reduction—myrcene, limonene, linalool, terpinol
  Relaxing—myrcene, limonene, terpinol, terpinolene, phytol

- **THCA**
  Reduces nausea and vomiting
  Neuroprotective
  Anti-inflammatory effects

Full-Spectrum Cannabis

Cannabis is a polypharmaceutical herb—a long word to describe the fact that cannabis contains a great variety of biologically active ingredients that can be harnessed to produce specific therapeutic effects relevant to hundreds of different patient populations. Another and perhaps more common term used is full-spectrum cannabis—the whole flower or plant, not just one or a few of its ingredients.

Starting more than a decade ago, a number of researchers posited, for two reasons, that applying a polypharmaceutical version of cannabis (e.g., flower-derived) works better than single-ingredient drugs (e.g., Marinol). One, the potential adverse effects of one ingredient can be mitigated by another, and two, multiple active ingredients can increase therapeutic efficacy.

Recent studies have since demonstrated and confirmed the earlier hypothesis that while isolated plant components can produce valuable therapeutic effects all by themselves, a full-spectrum medicine can reach deeper, be safer, and produce longer-lasting effects, while using less in comparison.

And while some of the workings of full-spectrum cannabis have been elucidated, the complete and precise mechanisms by which cannabis produces complex and synergistic effects are subject to ongoing investigations. These include studies of the interactions between the plant's native individual constituents such as cannabinoids (>100), their acid forms, terpenoids/terpenes (>100), hydrocarbons (>50), cannaflavins (>20 flavonoids), fatty acids (>30 mainly unsaturated fatty acids such as linoleic acid in seeds), carbohydrates (mono-, di-, and poly-saccharides), nitrogen-containing compounds including proteins, a number of other compounds (e.g., sterols, alcohol, esters), and one vitamin (K).

Key Points, Step 5
- Researchers posit that, based on current analytical limits and technological measuring capabilities, a threshold of **terpene concentrations of 0.05%** or higher is necessary to expect pharmacological effects in humans.
- Match cannabinoid-induced effects with similar terpene-induced effects to enhance therapeutic activity.
- Use terpene-induced effects to reduce a specific cannabinoid effect but still get the benefits of a high dose.

Step 6: Your Experience

In this section you will learn about the following:
- Setting a conscious intention for what you want
- Creating the ideal space for your cannabis experience
- Being with the experience
- Why and how to keep a session log

Your experience is perhaps the most important step you can take to discern the optimal plant constituents for what ails you or for the qualities of well-being that you desire. In other words, becoming aware of how and which plant ingredients individually but especially in concert affect your body, mind, and emotion are an indispensable opportunity to dial in those plant constituents that will work best for you.

Noticing and recording subtle or not so subtle changes will allow you to become aware of your body's as well as your mind's responses in general but, perhaps more importantly, to the specific changes you are making at any given step of the CannaKeys Process. Keeping a record of your experiences allows you to fine-tune the results. Pay attention especially to four areas of measuring changes or progress: symptoms reduction, changes in cognition, mood improvements, and the occurance of any potential adverse or unwanted effects.

At the end of this section, we have provided a number of worksheets you can use to record your feedback in such as a way as to easily make the adjustments needed to optimize your medicine. Each section on the session log is a reminder of the five steps described earlier in this chapter. And this (step 6) is the concluding step of the CannaKeys Process.

The Intention

To be clear, intention is not wishful thinking, which tends to be broad, nebulous, or abstract in comparison and thus lacks the clarity and focus you need to realize and sustain the shifts and changes you are looking for. In fact, research from mind-body medicine has shown that setting a direct, clear, and focused intention, especially when it is supported by an unwavering commitment, can activate and support the body's capacity for self-healing and resilience.

To begin, you may want to set an intention of what you want to focus on. It may simply be a symptom reduction or an improvement in your mood. However, you can also use your cannabis experience to assist you in creating the growth you would like to realize in yourself (e.g., changing unhealthy choices, beliefs, behaviors, healing knowledge you refuse to learn). If so, you may be setting the intention to get free from old patterns or stories that have become a liability. It may be looking for the inner resources to address a particularly difficult emotion such as shame, despair, or abject loneliness. It may be just to loosen up. You may want to allow a whole tapestry of emotions to be your guide to deeper self-knowledge. After all, self-respect comes from honoring all of your emotions. You may want to focus on why a difficult-to-be-with person acts the way they do. You may want to set an intention of releasing any sense of separation and replacing it with a profound sense of connection. You may want to seek deeper friendship, a sense of spiritual community, or a more intimate and lasting relationship. And it may come as a surprise that many times these non-physical changes also make a real and positive difference in realizing the symptom reduction and mood improvements you are looking for.

You may want to begin by deciding: "I want to heal" or "I want to heal, even if I don't know how" or "I want to discover what is blocking my healing." Or you may want to ask a specific question relevant to your specific symptoms.

Since people are unique in both biology and internal architecture of mind, we each have our own set of questions to explore, including those raised by the crisis of a mild, acute, or debilitating illness. Indeed, in the years of doing research and writing about integrative approaches to health and healing, I have come to look at symptoms as an expression of the body's innate intelligence, and that depending on severity, they are either an invitation or a demand for change. The innate part of you that is looking for healing has been trying to get your attention for a while. It will respond.

The Space

After you have chosen what to explore, move to create a safe and pleasant space for your experience. Turn off or mute your communication devices for the next few hours to make a pleasant and gentle environment. You may want to invite a trusted friend to hold space and provide support, empathy, and compassion or just hold the same space for yourself. If you want, use recordings of nature sounds or music you find especially inspiring. Draw the curtains, turn off or dim the lights, use an essential oil, rich in your favorite terpene in a diffuser, or put a drop or two on your wrist. Alternatively, a lot of people wishing to look inward opt to use eyeshades so as not to be distracted by external influences.

By now you likely have completed the previous five steps and have made an informed choice about what type, ratio, form, dose, and other cannabis constituents you are going to use. Now is the time for your cannabis experience.

The Experience

Relax, breathe, go with the flow, be with what surfaces. If you are experiencing tension, pain, unpleasant sensations, go with it. On the other hand, if your experience takes you into a different direction such as experiencing pleasant, sensual, or euphoric emotions, go with that as well. Both directions of emotional realities may be needed or hold insights relevant to your healing process. Trust your journey.

Like many others who have used this method for deep healing, you may become (more) aware of some internal resistance to getting better, which could be any behavior that propels your attention outward, such as paying attention to noise elsewhere, wanting to talk to someone, or wanting to move around. Signs that relevant material is emerging may include feeling uncomfortable, distracting pain, or falling asleep. Try to stay focused within. Notice your sensations, desires, and impulses; be curious and try to discover what lies beneath. Under the influence of cannabis or through engaging a spirit of gentle mindfulness, you may find, to your surprise, that you are fully able and willing to forgive yourself for any resistance present. This is an excellent start.

However, once you are past the peak experience as indicated by returning to a more normal state of consciousness, bring in your initial intention. This is a good time to engage and to merge your intention with the flow of the experience. Focus on your primary symptom, focus on your intention and notice what is happening in specific tissues of your body, in your body at large, notice your sensations, feeling, emotions, quality of thought or self-talk.

In particular, many people finding it instructive and helpful to focus on how you feel about what your symptom is preventing you from doing. It may help to think of these feelings as the needle of a compass that points to underlying issues that require your attention. Often, at this point of the process, patients report the emergence of a memory or a related scenario. Sometimes several memories or scenarios emerge, and, not unlike layers of an onion, they produce tears or reveal an emotional connection or similarity, and thus a theme that may be important to notice on your journey to health and well-being.

You may discover that merely being with whatever emotions or resistances surface, no matter how unpleasant or difficult it at first seems, is the beginning of true, deep healing. Once you have released enough of your emerging emotional material, you may be able to identify relevant aspects of mind such as beliefs, attitudes, and choices that are contributing to your ill health and which are in need of replacement. These would be good to notice and remember.

Here are some sample questions you may want to consider to gain insight and direction about how best to proceed.

- What is this condition taking from me?
- What is it giving me?
- What are the symptoms keeping me from doing?
- How do I feel about that?
- Where do these feelings take me?
- What scenarios or memories are associated with these feelings?
- What is the unhealthy meaning I have given to that scenario?
- How can the answers to these questions support my capacity for self-healing?

The Session Log

What is the best way to keep a record of your experiences? Consider how you best remember things or events. When remembering, do you tell yourself what happened? Do you see it as a picture or a movie in your mind's eye? Is it a feeling that conveys and communicates your memory? Is it a combination, and if so, which sense tends to stand out?

Once you have a pretty good idea of how you remember, it is helpful to find the easiest or most elegant way to record it for easy access and comparison later. Is it a series of keywords to remind you? Is it a number of bullet points highlighting key elements of your experiences? Is it an image or a mental snapshot combined with describing some details that are important to you? Is it writing a short or long narrative? Or is it something in between? The answer depends on you and your preferences. Whatever style, you may find it helpful to focus on symptom reduction, cognitive shifts, changes in mood and emotion, and adverse effects.

Symptom Reduction

- Consider starting by choosing one symptom you want to reduce or eliminate. If you are dealing with more than one, make a list in order of priorities and start at the top.
- What do you notice after consuming cannabis? Is it the same, better, gone, or worse?
- If you feel better and you are happy with the results, stop. You have found your optimal plant remedy for where you are now in your healing process (which may shift slightly over time). Use the same chemotype, ratio, form, dosage, other plant ingredients/ratios the next time you need it. (When you go to a dispensary to refill your prescription, choose one that has the same numbers, which may or may not be the same strain you purchased for this experience).
- If adverse effects occur, stop. Wait a day, look at your notes; make one change in the CannaKeys Process; and start again the next day.
- If the same, and you experience no adverse effects, consider making one change in the CannaKeys Process and try again.
- If inhaling, note the number and depth of inhalation (e.g., shallow, medium, or deep) in your journal.

Sometimes the difference in symptoms is subtle but still present, so a good way to track subtleties is using a scale. For instance, hospitals use a scale for describing pain between 1 and 10, with 10 being the worst pain you have ever experienced and 1 being very tolerable. You can use the scale for any specific symptom to monitor subtle progress and continue to fine-tune your remedy. Simply cross out pain and write in the symptom you want to measure.

Pain Measurement Scale

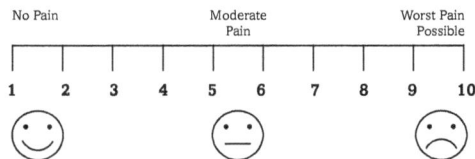

Chapter 5, Graphic 4: Pain Measurement Scale

Did you notice any positive, expansive, pleasant sensations (e.g., softness, warmth, coolness, smoothness of being, gentleness of touch)? In the absence of adverse effects or lack of progress with symptom reduction, these are positive feedback sensation that signal progress. Note these sensations in your journal as well.

Changes in Cognition

The word *cognition* is based on the Latin word *cognitio*, meaning "acquiring knowledge." Cognition is the process of reaching for understanding through thinking. You may be surprised to learn that most of us have an innate human opposition to acquiring knowledge, which is referred to as status quo bias. We are creatures of habit, we like the way things are even if sometimes it is less than optimal or even detrimental. One of the reasons I have come to believe cannabis works so well for so many different patient populations is that it makes it easy and fun to gently push through our natural resistances to change. And while our mind is relaxed and busy finding funny things in the mundane, the plant engages our deeper biology via various ECS-based mechanisms to bring about and maintain system-wide homeostasis.

You may want to think of it as follows: Contributing factors for many conditions, especially those of a chronic nature, rarely exist in the physical realm alone; they tend to have mental and emotional correlates. For instance, the mind-set that made you vulnerable to a particular disease expression is not the mind-set that is going to get you out of it. So a gentle, deeply relaxing, revelatory shift in cognition may be one of the most important reasons why cannabis is helping so many. What should you pay attention to? What do you write down?

Start by noticing the quality of your thoughts. A good way to notice your thoughts is by listening to your self-talk. How fast is that inner voice? (Speed often is associated with fear or anxiety on one end of the spectrum and excitement on the other.) What is the sound of your inner voice? Does it have a high pitch or a low bass-like tone, or is it something in between? Are your thoughts harsh? Are they soft? Are they punishing or nurturing? Are they supportive or judgmental? You may find it easier to recognize the quality of your voice by the quality of emotions that tend to follow each internal sentence completed. Are you anxious, frustrated, relaxed, elated, peaceful, tense, relaxed? Make a note of what you notice before, during, and after the cannabis experience.

Do you notice any patterns in your thinking? Do you notice triggers that when experienced set the same response into motion? If so, how does the cannabis experience shift the pattern or triggers? Does the experience allow you to be in the presence of otherwise intolerable emotions? What does the experience reveal to you about you? For instance, did you notice any limiting beliefs or disempowering perspectives? Did you perceive new ways of letting them go and replacing them with a new sense of what is possible? Is what you are learning helping you in identifying and using hitherto-unknown inner resources?

Mood Improvements

Consider starting by choosing one mood you would like to improve upon. If you would like to work on others, make a list, prioritize them, and start at the top.

What happened to the specific mood improvements you attempted to address? Are you on target? Do you feel better, lighter, more relaxed? Do you feel more resilient, supported, connected to nature, friends, family, lover(s), or a celestial architect of your choice?

Do you experience a constricting shift in mood (e.g., anxious, depressed, scared, gloomy, dreading, panicky)? If yes, consider reducing THC content and/or increasing your CBD content. Make one change in the CannaKeys Process and take note of it in the journal.

Do you experience an expansive shift in mood (e.g., uplifted, relaxed, happy, euphoric, mirthful, laughing, blissful)? If yes and you like it, stay the course. If not, make one change in the CannaKeys Process. By now you know your options (change chemotype, ratio, form, dosage, entourage).

Adverse Effects

Notice any unwanted effect. What do you notice? Any new and unwanted sensations such as an increase in any chronic symptoms including pain, nausea, vomiting, fear, anxiety, panic? What about changes in physical functions such as an unsteady gait, difficulty talking, or confusion? Stop and consider lowering THC intake by reducing total amount, switching to a lower potency or ratio, or going to chemotype III.

Session Log (to review and adjust, make changes one step at a time):

Primary symptoms or quality of well-being you wish to improve:

_____

Secondary symptoms or quality of well-being you wish to improve:

_____

Step 1: Record the chemotype (CT) of your choice:

☐ CT I (More THC less CBD)   ☐ CT II (Equal THC:CBD)

☐ CT III (Less THC more CBD)

Step 2: Record your specific THC:CBD sub-ratio _____:_____

Step 3: Record your preferred form

☐ Inhalation   ☐ Ingestion

☐ Other: _____

Step 4: Record amount used mg of THC and CBD

THC _____mg CBD_____mg

Step 5: List other plant ingredients (i.e. notable terpenes, other cannabinoids, acid forms)

_____

Record your responses:

_____

_____

Before treatment: After treatment:

Pain Measurement Scale   **Pain Measurement Scale**

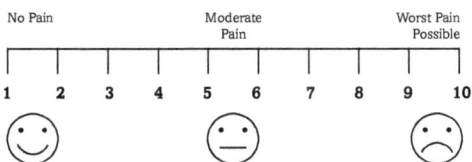

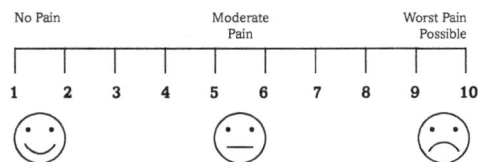

No Pain	Moderate Pain	Worst Pain Possible		No Pain	Moderate Pain	Worst Pain Possible
1 2 3 4 5 6 7 8 9 10				1 2 3 4 5 6 7 8 9 10		

Session Log (to review and adjust, make changes one step at a time):

Primary symptoms or quality of well-being you wish to improve:

_____

Secondary symptoms or quality of well-being you wish to improve:

_____

Step 1: Record the chemotype (CT) of your choice:

☐   CT I (More THC less CBD)   ☐   CT II (Equal THC:CBD)

☐   CT III (Less THC more CBD)

Step 2: Record your specific THC:CBD sub-ratio _____:_____

Step 3: Record your preferred form

☐   Inhalation   ☐   Ingestion

☐   Other: _____

Step 4: Record amount used mg of THC and CBD

THC _____mg CBD_____mg

Step 5: List other plant ingredients (i.e. notable terpenes, other cannabinoids, acid forms)

_____

Record your responses:

_____

_____

Before treatment: After treatment:

Pain Measurement Scale   **Pain Measurement Scale**

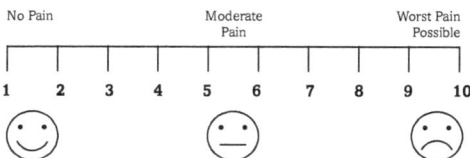

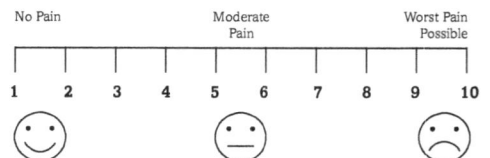

No Pain   Moderate Pain   Worst Pain Possible   No Pain   Moderate Pain   Worst Pain Possible

1 2 3 4 5 6 7 8 9 10   1 2 3 4 5 6 7 8 9 10

Session Log (to review and adjust, make changes one step at a time):

Primary symptoms or quality of well-being you wish to improve:

_____

Secondary symptoms or quality of well-being you wish to improve:

_____

Step 1: Record the chemotype (CT) of your choice:

☐   CT I (More THC less CBD)   ☐   CT II (Equal THC:CBD)

☐   CT III (Less THC more CBD)

Step 2: Record your specific THC:CBD sub-ratio _____:_____

Step 3: Record your preferred form

☐   Inhalation   ☐   Ingestion

☐   Other: _____

Step 4: Record amount used mg of THC and CBD

THC _____mg CBD_____mg

Step 5: List other plant ingredients (i.e. notable terpenes, other cannabinoids, acid forms)

_____

Record your responses:

_____

_____

Before treatment: After treatment:

Pain Measurement Scale   **Pain Measurement Scale**

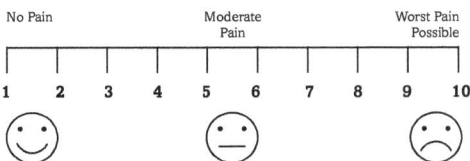

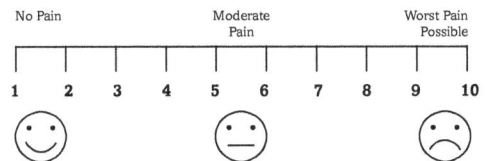

| No Pain | Moderate Pain | Worst Pain Possible | | No Pain | Moderate Pain | Worst Pain Possible |

1 2 3 4 5 6 7 8 9 10   1 2 3 4 5 6 7 8 9 10

☺   😐   ☹   ☺   😐   ☹

Session Log (to review and adjust, make changes one step at a time):

Primary symptoms or quality of well-being you wish to improve:

_____

Secondary symptoms or quality of well-being you wish to improve:

_____

Step 1: Record the chemotype (CT) of your choice:

☐   CT I (More THC less CBD)   ☐   CT II (Equal THC:CBD)

☐   CT III (Less THC more CBD)

Step 2: Record your specific THC:CBD sub-ratio _____:_____

Step 3: Record your preferred form

☐   Inhalation   ☐   Ingestion

☐   Other: _____

Step 4: Record amount used mg of THC and CBD

THC _____mg CBD_____mg

Step 5: List other plant ingredients (i.e. notable terpenes, other cannabinoids, acid forms)

_____

Record your responses:

_____

_____

Before treatment: After treatment:

Pain Measurement Scale   **Pain Measurement Scale**

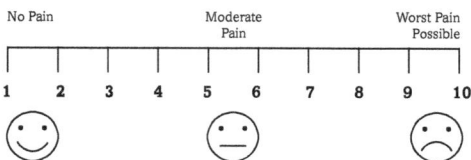

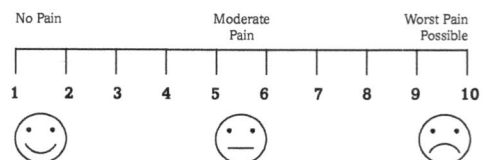

No Pain   Moderate Pain   Worst Pain Possible   No Pain   Moderate Pain   Worst Pain Possible

1   2   3   4   5   6   7   8   9   10   1   2   3   4   5   6   7   8   9   10

Session Log (to review and adjust, make changes one step at a time):

Primary symptoms or quality of well-being you wish to improve:

_____

Secondary symptoms or quality of well-being you wish to improve:

_____

Step 1: Record the chemotype (CT) of your choice:

☐   CT I (More THC less CBD)   ☐   CT II (Equal THC:CBD)

☐   CT III (Less THC more CBD)

Step 2: Record your specific THC:CBD sub-ratio _____ : _____

Step 3: Record your preferred form

☐   Inhalation   ☐   Ingestion

☐   Other: _____

Step 4: Record amount used mg of THC and CBD

THC _____ mg CBD _____ mg

Step 5: List other plant ingredients (i.e. notable terpenes, other cannabinoids, acid forms)

_____

Record your responses:

_____

_____

Before treatment: After treatment:

Pain Measurement Scale   **Pain Measurement Scale**

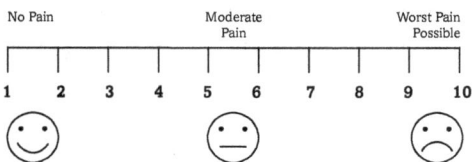

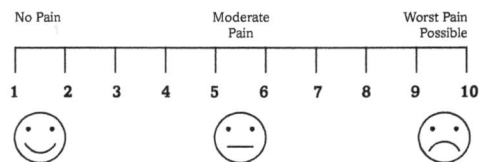

No Pain	Moderate Pain	Worst Pain Possible		No Pain	Moderate Pain	Worst Pain Possible
1 2 3 4 5 6 7 8 9 10				1 2 3 4 5 6 7 8 9 10		

Key Points, Step 6

Set a conscious intention

Make some changes to your environment to support your experience

Go with the flow of experience until it is just past its peak before you focus on a more conscious exploration of your healing process

When it comes to dialing in your optimal cannabis composition, make one adjustment at a time and carefully record your experience with each change at each step.

To remember important details, it helps to write down your experience in your own session log or use the worksheets in chapter 5.

Consider focusing on the following:
- Shifts or changes in the symptom you are treating
- Shifts and changes in the qualities of being you wish to improve upon
- Shifts or changes in cognition; track the quality (e.g., speed) of your thoughts and self-talk (e.g., negative, nurturing)
- Shifts or changes in quality of mood, feeling, emotion
- Presence or absence of adverse effects

Concluding Thoughts

Patients, especially those with chronic conditions, know very well that the orthodox medical approach is focused on finding one or a number of pharmaceutical "silver bullet(s)"—a quick shot to take away the pain, a weapon to diminish or demolish the anxiety, or to deal a powerful blow to alleviate depression. However, as you very well know, gunfights can cause collateral damage. Bullets that temporarily take away the pain, anguish, and hopelessness now produce addiction and death by overdose in record numbers.

The good news is that a brighter future world of possibility is emerging. In numerous countries around the globe, in states and cities across the US where cannabis is available, large numbers of patients are switching to cannabis. Cannabis reduces pain, anxiety, and depression, but it does so as an embrace, not as a bullet. In fact, I have come to believe that cannabis will never be a silver bullet. Why?

Cannabis can show each of us what our unique map to healing looks like. It is a guide that will not replace our healing work but can show us ways to make our healing journey easier, safer, and more effective. Here are a few examples.

Research has shown that chronic negativity (such as excessive fears, anxiety, shame, stress) lowers our immunity and resilience[28,29] and like a virus spreads internally and is contagious externally. How can cannabis help? The plant is broadly recognized for its capacity to diminish negativity and replace it with a gentle attitude, an easy smile, and optimistic outlook. . . . Isn't that often why we use it?

Second, science tells us that suppressed or repressed emotions are a disease. For instance, consciousness research in cancer has discovered that the single most common mental-emotional trait that determined cancer survival was a sense of emotional authenticity.[30] How is cannabis helpful? Many cannabis-using patients dealing with various cancers report disease regression. Why? This could be partly because cannabis offers the deep relaxation and frame of mind that makes it easier to accept oneself, especially in the presence of otherwise intolerable feelings or emotions.

For instance, the Cornell Medical Center reported a case of a patient suffering from chronic and severe hypertension with complications that could not be controlled, even with aggressive pharmaceutical interventions. It was not until she discovered incest-related emotional material and worked through it that a cure was within reach. When she was done with that layer of her healing work, her hypertension disappeared.[31]

As a third example, recent studies demonstrate that forgiveness has a powerful ability to support our healing process.[32,33] I don't know about you, but for some of us forgiveness is not that easy. . . . How could cannabis be helpful? Cannabis has the capacity to send a pink slip to the slave driver in the back of our mind that insists on punishment and blame as tools for growth and motivation. As a result, implementing forgiveness becomes easier and more elegant, and in doing so contributes to our healing.

Whether we are choosing to work with cannabis as an ally, whether we are working with mind-body medicine or mindfulness alone, or whether we use a combination, the more we become conscious about these natural mechanisms to deepen our healing process, the more our natural healing abilities emerge and increase.

When we relax and relax deeply, we enter an altered state of consciousness where it becomes easier to engage our emotional intelligence to release repressed emotions and elevate the way we feel. As a result, the body's entire neurochemistry will respond in kind.

In many cases that may just be enough to unblock our capacity for healing, or at the very least, it is a good beginning.

The Science Behind the Top 10

- Cancer
- Heart Disease
- Chronic Pain
- Diabetes
- Anxiety
- Neurological Conditions
- Insomnia
- Colitis
- Cough
- Skin Conditions

Cancer

There are over a hundred different types of brain cancers. One of the more common and malignant types are gliomas. The word *glioma* is compiled from the Greek *glia* "glue" and the suffix *-oma,* meaning "morbid growth or tumor." This type of cancer tends to form in supportive cells (glial cells) of the brain or spinal cord. Gliomas are defined or categorized by the various types of glioma cells in which the cancer originated (e.g., subtypes include astrocytomas, ependymomas, or oligodendrogliomas). The location and size of cancerous brain tumors largely determine survivability and the various signs and symptoms likely to develop. As tumors grow, impairment increases. As a result, the whole body as well as the mind may be affected. Generally speaking, the lower in the brain structure the tumor is, the poorer the survival outcome. This is because the brain's lower portions control most vital functions such as breathing and heart rate.

Signs and symptoms cover a relatively wide range and can include altered levels of consciousness ranging from mild confusion to epileptic seizures, from odd behavior to full-blown

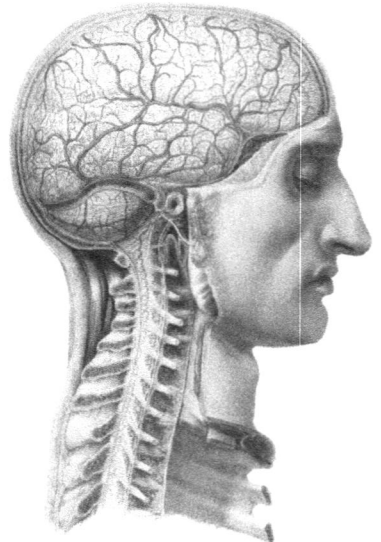

Appendix, Graphic 1

stroke-like handicaps. Additional commonly reported symptoms include headache, visual impairments, and nausea with vomiting.

Cannabinoid-Based Research Highlights Relevant to Brain Cancer

While a number of cases of significant improvement in patients with a very poor prognosis of glioma have been described, the exact mechanism by which cannabinoids produce their therapeutic effects remain to be elucidated.[1] Of the studies reviewed for this section, all but two were limited to laboratory, animal experiments, and reviews of the currently available literature provide us with somewhat limited experiences to use to make more informed decisions.

However, these pre-clinical trials do reveal numerous cannabinoids (endo-, phyto-, and synthetic) that activate a number of biological pathways which exert potential therapeutic effects. To look at it in a relatively chronological fashion, the earliest evidence for beneficial endocannabinoid system involvement in cases of brain cancer came in 1998, when European researchers discovered that THC was able to instruct cancerous glioma cells to commit suicide (apoptosis) by a mechanism not related to CB1 receptor sites.[2]

The next clue came three years later. An animal trial in 2001 conducted by Spanish researchers discovered that the injection of a synthetic cannabinoid (JWH-133, a potent CB2 agonist) directly into the affected tissue of mice induced a considerable regression of malignant glioma cells, and thus suggested a role for CB2 receptor sites in producing therapeutic effects.[3] (The reader is reminded that the brain and spinal cord primarily contain CB1 receptor sites. However, these researchers found CB2 receptor sites expressed in the cell membranes of gliomas and were able to produce the described biological effects. If these results are confirmed in clinical trial conducted on humans, it would potentially allow for treatment without changes in cognition.)

Shortly after these findings two additional studies from Italy expanded our understanding of brain cancer and cannabinoids by suggesting that non-psychoactive CBD was able to produce a significant anti-brain tumor activity, both in vitro and in vivo,[4] and that CBD selectively produced oxidative stress in brain cancer cells, thus producing apoptosis, leaving normal cells unaffected.[5]

In 2004 another Spanish experiment conducted on mice and on two human patients with gliomas produced another perspective as to why cannabinoids may present a new therapy for patients suffering from brain cancer. Results showed that cannabinoids effectively inhibited a chemical signal needed for the brain tumor to build its blood supply, an essential element for its survival and proliferation. The Spanish researchers considered the blockade of this signal to be one of the most promising anti-tumor approaches currently available, and the work proposed "a novel pharmacological target for cannabinoid-based therapies."[6]

By 2005 the anti-tumor properties of CBD emerged as a topic of exploration in an experiment that demonstrated that the cannabis-based compound was able to inhibit the migration of glioblastoma cells (U87) by a mechanism independent of CB1 or CB2.[7]

In 2006 Italian scientists describe another potential therapeutic ability of CBD to induce apoptosis of glioma cells and tumor regression by a mechanism involving the activation of enzymes (e.g., caspase-3) and reactive oxygen species.[8] Indeed, CBD seems to counteract glioma cell migration, proliferation, and invasion by a number of yet-to-be-understood mechanisms.[9]

That same year neighboring Spanish physicians conducted the first human trial on nine patients with recurring forms of glioblastoma multiforme who did not respond to conventional treatments and whose tumors were demonstrably spreading. Each had THC directly injected into their tumors to determine if it was safe and to see what impact it had on affected glioblastoma cells. Results showed that "Delta(9)-Tetrahydrocannabinol inhibited tumour-cell proliferation in vitro. The fair safety profile of THC, together with its possible antiproliferative action on tumour cells reported here and in other studies, may set the basis for future trials aimed at evaluating the potential antitumoral activity of cannabinoids."[10]

Since then, a number of additional studies discovered novel cannabinoid-based action that may produce therapeutic effects against brain cancers. For instance, THC was shown to inhibit two proteins that glioma cells thrive on.[11] THC (as well as JWH-133, a potent CB2 agonist) also downregulated tissue inhibitors of metalloproteinases (TIMPs), which enhances invasive capacities of tumor cells.[12] In another study, a team of international researchers discovered another potential mechanism by which THC kills brain cancer cells. THC increases the dihydroceramide:ceramide ratio in glioma cells, which led to the alteration of certain aspects of the cellular architecture and molecular mechanisms, thus promoting death of brain cancer cells.[13]

Scientists from San Francisco's California Pacific Medical Center conducted a laboratory experiment and demonstrated in part that two cannabinoids, THC and CBD, acted synergistically to inhibit brain cancer cell growth by inducing reactive oxygen species to produce apoptosis.[14]

Discuss with your licensed and cannabis-experienced physician if the following considerations for specific chemotypes, sub-ratios, forms, dosage, terpenes, or other plant constituents as well as their respective summaries are safe and of potential therapeutic value for you.

Chemotype-Based Specific Considerations for Brain Cancer

Both THC and CBD as well as a number of synthetic cannabinoids (e.g., JWH-133) and endocannabinoids (e.g., anandamide) have been found to be potent inhibitors of brain cancer cell development and to be able to induce apoptosis (cancer cell death). Each works by a unique mechanism not limited by activation of the classical endocannabinoid receptors CB1 or CB2. Greater effects may be produced synergistically when THC and CBD are applied together. Thus all three chemotypes may show promise. Since each chemotype may show degrees of efficacy, you can use your preferences about what kind of cannabis experience you wish to have as a way to inform your choices. If you wish to avoid cognitive changes, if you wish to avoid any potential adverse effect (within normal dose ranges), and if you are concerned about the relatively small addiction potential, you should start with a chemoype III with less than 0.3mg of THC to meet all your preferences and still be able to tap into the plant's potential therapeutic abilities.

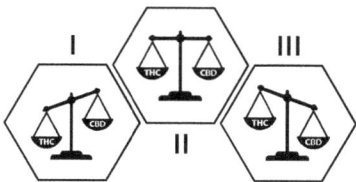

Sub-Ratio

Since all three chemotypes may provide support, sub-ratios too can be informed by your preferences.

Form Considerations

Inhalation can provide you with near instant results that tend to last for up to two hours, while an ingestible cannabis-containing product will take longer (up to two hours) to take effect but will last for longer periods of time (e.g., four to eight hours). Using rectal suppositories containing cannabinoids has been reported anecdotally effective in a number of individual case studies including glioma. And, while scientific data about rectal applications is still very limited, the relatively high safety profile combined with traditional usage reports as well as case-based trends make this form a potentially attractive and novel (or perhaps not so novel) option especially for patients who are more sensitive to the effects of THC or those who wish to minimize potential psychotropic shifts or changes.

Dosage Consideration

The optimal dosage depends on a number of factors. Are you looking to treat a particular symptom such as a cancer-induced pain, are you treating a chemotherapy- or radiation-induced symptom such as nausea or vomiting, or are you trying to employ the potential anti-cancer properties for a deeper approach?

To date, no clinical or side-by-side studies have been conducted to examine dosing details in the context of treating cancer (or cancer-related symptoms), and thus a practical and actionable scientific evidence–based approach is still wanting. Lack of scientific evidence is the main reason a lot of physicians are unwilling to advise patients on the use of cannabis in the treatment of cancer-related symptoms or cancer itself.

However, based on a number of successful cases that I have been following, a symptom treatment approach tends to involve low to medium dose ranges between 5 to 20mg THC.

In contrast, deep-healing approaches utilizing the potential apoptotic effects (cancer destroying) of cannabis may require higher dosages. Similarly, in a number of cases that I have been tracking where patients went into full remission even with end-stage types of cancers have been using primarily a concentrated cannabis oil (think reduction source).

There are a great number of product makers and purveyors, and most of them have their own way of making the oil. However, no matter what specific method is used, they all are basically an alcohol or solvent reduction that leaves a tar-like oil abundant in THC and other plant constituents, with water evaporated and solvents reduced to trace amounts. Reductions may vary in the source materials used and may include high-proof grain alcohol (e.g., Everclear 95% alcohol by volume), pure light naphtha (a controversial petroleum product not available in the US), or food-grade isopropyl alcohol (99.9%). THC percentages can vary somewhere between 50 and 90 percent. It may seem counterintuitive, but plant matter soaked longer tends to contain less THC when compared to shorter soak times of around five minutes. The reduction or tar-like oil is usually drawn up into a needle-less syringe for ease of usage, to store, and to dispense, which is somewhat troublesome to do when fluctuations in temperature produce changes in consistencies. The cooler it is, the more stiff and solid-like the oil will become. With warmer temperatures the oil can be very smooth.

Over time, various advocates, patients, and their caregivers have coalesced to propose a three-month regimen with the aim of using up to 60,000 mg (60 gm) of the oil over that period of time. In their treatment regimen, proposed daily dosages build up gradually to an extremely high dosage of up to 1,000 mg of the concentrate. However, remember the actual ingested THC amounts depends on the concentration of THC in the oil reduction. For instance, 1,000 mg of the oil reduction with a 70% or 90% THC concentrate would yield 700 mg THC or 900 mg THC, respectively (e.g., 1,000 mg oil × 70%THC ÷ 100% = 700 mg THC). The remaining 300 mg (or 100 mg) are other plant constituents such as terpenes or chlorophyll.

To make these proposed dosing procedures more practical, some visual comparisons are commonly used (e.g., the single small grains of rice) and some assumptions are made such as 1,000 mg (1 gm) is about equal to 1 mL of the reduction/oil.

To let the body get accustomed to these very potent applications, many patients use a "buildup" regimen to make their cannabis experience the best that it can be. It begins with using a tiny amount of the sticky (tar-like) oil or about one quarter of the size of a dry short grain of rice. The amounts are gradually increased until about 1,000 mg of the reduction is used daily. The total daily amount is divided into three equal dosages about eight hours apart (am, noon, and pm) of approximately 333 mg of the oil or roughly the size of ~15 to 20 single small grains of rice. After the regimen is completed, many patients opt for a maintenance program that is significantly reduced in amounts taken (~1 gm a month).

Alternatively, a number of patients who wish to employ high dosages of THC but who prefer a "no-high" experience use the same regimen but as a suppository. The majority of the patients I interviewed who used highly concentrated cannabis suppositories reported no or only a very limited experience of feeling high. However, I know of at least one case where the patient felt "high" for about 16 hours. Once more, experience demonstrates that cannabis is a subjective medicine that breaks the mold of one size fits all.

Take Note: To date, no clinical trials have examined the effectiveness of high-dosage approaches on gliomas (or other types of cancers). Anecdotal reports are commonly used by proponents of high-dose THC oils to describe its potential effectiveness to work in "most cases." However, extensive interviews with cannabis-experienced clinicians suggest two diverging views. One, about one half of all cancer patients using high-THC-dose regimens fail to respond (which is still a good portion), and two, a number of patients respond really well using lower-THC-dose approaches.

Designing an Entourage Effect for Cancer

There are a number of terpenes that have demonstrated potential for destroying cancerous cells, such as eucalyptol for leukemia[15]; limonene in cases of breast[16]; linalool for kidney,[17] breast,[18] lymphoma, and leukemia type cancers;[19] pinene for liver cancer[20]; terpineol for cancer of the lungs[21] and leukemia[22]; beta-caryophyllene against colon and pancreatic,[23] melanoma and lymphoma,[24] and kidney cancer[25].

Evidence for the use of terpenes in cases of brain cancer point to a naturally occurring monoterpene called perillyl alcohol (POH) and its precursor limonene.[26] While ingesting POH may be prohibitive due to gastric toxicities, clinical trials from Brazil using inhaled forms suggest a beneficial role, free of toxic effects, and regression of tumor size in some patients.[27] These results were confirmed by another trial from Brazil conducted on 89 patients suffering from glioblastoma. Roughly half of the group of patients received 440 mg of POH (administered intra-nasally) while the other half did not. Those treated with POH lived significantly longer than the untreated group.[28]

Possible Synergistic Effects of Cannabinoids and Pharmaceutical Drugs

Beyond that of THC/CBD-based synergies, research results also demonstrated that cannabinoids increase the efficacy of pharmacological chemotherapies such as temozolomide,[29] carmustine, or doxorubicin[30] and gamma radiation (CBD).[31]

Adverse Effects and Contraindications with Pharmaceutical Drugs

To date, there has been a limited amount of research into the interactions between cannabinoid and chemotherapeutic agents (or radiation treatments). However, out of the limited data currently available, a number of relevant trends have emerged that are worth considering. One, widespread observational cases have shown that a full spectrum usage of cannabis in its acid form (non-decarboxylated) does not negatively affect orthodox cancer treatments. Two, evidence from human trials has shown that co-administration with cannabinoids can mitigate adverse effects of chemotherapy (e.g., chemotherapy-induced nausea/vomiting). Three, a growing number of trials are also examining the potential apoptotic (cancer killing) effects of cannabis itself. Four, studies indicate potential synergistic effects of cannabinoids that support the potential therapeutic effects (which may require a reduction of chemotherapeutic agents). Five, early studies indicate that cannabinoids such as CBD or THC interact in a variety of ways (e.g., inhibit, block, enhance) with a family of enzymes (Cytochrome 450) that break down a majority of all pharmaceutical drugs on the market. High dosages of an isolated cannabinoid may make a pharmaceutical drug (e.g., chemotherapy) less effective and/or more toxic.

Summary: Cancer

Step 1 Chemotype: Current evidence suggests possible therapeutic effects with all cannabis chemotypes (I, II, and III).

Step 2 THC:CBD Ratio: If you are treating symptoms and are concerned about psychotropic effects, consider starting with those ratios lowest in THC. For chemotype-I (consider starting carefully with an equal to or less than 5:1 ratio); for chemotype-II, start with the lowest available level of THC (1-4%). If results fail to mitigate your symptoms, consider gradually increasing THC content.

Step 3 Choose Preferred Form of Inhalation: Consider smoking or vaporizing the cannabis flower to achieve quick and relatively short-lasting effects (about two hours). Consider an ingestible form when longer duration of effects is needed.

Step 4 Optimal Amount: Many cannabis-using patients who are ingesting, and who are looking for symptom relief, start with 5 mg THC and increase their dosage incrementally and slowly (leaving at least an hour in between).

Many times patients who are looking for deeper healing choose higher THC-dosage regimens.

Step 5 Design Entourage Effect: The naturally occurring monoterpene called perillyl alcohol (POH) and its precursor limonene, as well as beta-caryophyllene, limonene, linalool, pinene, terpinolene, have been described in the literature as potentially novel approaches.

Step 6: Keep careful notes on your progress and use the pain scale in chapter 5 to measure shifts in pain or any other specific symptom. Use the worksheet from chapter 5 to measure effects on your primary symptom before and after inhaling/ingesting. Revisit the scale at different times to better understand the subjective flow of effects in terms of duration.

Heart Disease

The heart is a unique muscle that gets its own nourishment such as oxygen from three main vessels called the coronary arteries. In Western medicine's model, when one or more of these coronary arteries is slowly or suddenly blocked (as in the case of a blood clot) or gradually narrows (as in the case of atherosclerosis, a buildup of plaque), it becomes deprived of oxygen, and the heart muscle begins to ache.

This process is called myocardial ischemia (an inadequate oxygen supply to the muscles of the heart). Depending on severity, ischemia can lead to irregular heartbeats (arrhythmias) that when severe enough or sustained over time can produce chest pain (angina) or progress to an infarct (heart attack), a potentially life-threatening condition.

However, merely restoring blood supply to the ischemic tissue may not solve the entirety of the problem. Actual practice shows that after emergency interventions (e.g., clot-dissolving drugs or surgery) the damage often increases.

This additional trauma to an already "aching heart" is referred to as an ischemia/reperfusion (I/R) injury. It is hypothesized that this damage is induced by free radicals or reactive oxygen species (ROS) such as superoxide (O_2-) or hydrogen peroxide (H_2O_2). The more ROS accumulates during a heart attack, the deeper the potential damage.

While healthy cells have the capacity to neutralize ROS by using enzymes (such as catalase for H_2O_2 or superoxide dismutase for O_2-) that disarm or detoxify them by changing them to harmless molecules such as regular oxygen or water, already stressed or damaged cells are not able to detoxify these radicals, thus creating additional compounding events such as the production of pro-inflammatory cytokines, toxic levels of the excitatory neurotransmitter glutamate, adhesion molecules, and immune reactions where the body's defenses begin to view the damaged cells as enemies and thus begin to attack them, causing more local damage as well as systemic harm.

In the bigger picture, while cannabinoid-based therapies cannot do anything about some of the contributing factors underlying the development of heart disease such as family history, quite a few of them may also respond positively to cannabinoids, for example chronic mental-emotional stress (including their physical correlates like inflammation-based stress and oxidative stress), hypertension, atherosclerosis, diabetes, and mood disorders (e.g., anxiety, depression).

Cannabinoid-Based Research Highlights Relevant to Heart Disease

First and foremost, the pre-clinical studies reviewed for this section showed numerous potential mechanisms by which cannabinoids modulate cardiac function.[1] Cannabinoids may present new mechanisms that may at once be cardioprotective and potentially therapeutically efficacious in dealing with contributing factors as well as numerous chronic and acute cardiac events including hypoxia (poor tissue perfusion), arrhythmias (irregular heartbeats), angina, myocardial infarction (heart attacks), and myocardial ischemia/reperfusion injury. Here is what the currently available evidence bears out so far.

One of the major contributing factors to cardiovascular illness is **chronic stress**. The right type of cannabis, at the proper dose, matched to the specific needs of each patient will produce a deep sense of relaxation and contribute to cardiovascular homeostasis. And, while a laboratory study[2] and a test conducted on mice[3] suggested cardioprotective effects for THC, researchers also noted as early as the seventies that THC also increases heart rates, thus suggesting limitations on its use in certain circumstances, such as in the presence of rapid heart rhythms (tachycardias) during a myocardial infarction.[4,5]

CBD has been shown to produce subtle therapeutic changes in actual hemodynamics such as changes in heart rate and blood pressure.[6] For instance, one study conducted on rodents discovered that CBD reduced all stress responses including that of a rapidly beating heart,[7] while another experiment conducted on healthy human volunteers discovered that a single dose of 600 mg of CBD was able to reduce blood pressure rise in response to stress.[8]

CBD was found to produce various **cardioprotective** effects via binding with PPARγ.[9,10] And, there is evidence to suggest that at least some of the cardioprotective effects of CBD are mediated by the modulation of certain anxiolytic neurotransmitter pathways (e.g., serotonin) that also affect hemodynamics such as blood pressure and heart rate.[11] In other words, CBD reduces anxiety during a stressful event by calming the body, mind, and emotion: no small task, since stress is a known contributing factor in the development of hypertension, atherosclerosis, and heart disease.

Consider that oxidation of LDL is known to be the first step in developing **atherosclerosis** and the enzyme 15-Lipoxygenase is believed to be involved in taking that first step.[12] Both CBD and a derivative CBDD (dimethyl ether) have been found to be potent inhibitors of this enzyme.[13] Another mechanism known to be pro-inflammatory and pro-atherogenic is activation of GPR55 receptor sites. THC is a potent GPR55 agonist, while CBD an antagonist,[14] which would make reducing THC and increasing CBD in such cases preferable.

An additional experiment conducted on mice showed that activation of CB2 receptor sites **reduces aortic atherosclerotic lesions** via reducing macrophage adhesion.[15] Further, researchers discovered a cannabis-based terpinoid able to bind with CB2, beta-caryophyllene (Ki ~ 155 nM), which introduced significant anti-inflammatory actions that might be relevant for patients with atherosclerosis.[16] If these results are confirmed in humans, beta-caryophyllene could be an important adjunct for patients whose cardiac conditions are exasperated by pro-inflammatory actions of the body's immune responses.

Additionally, trials conducted on rodents also showed that CBD exerts **cardioprotective** properties[17] in the acute phase of **ischemia/reperfusion injury**.[18,19] It was posited that the effects were realized by reducing **irregular heartbeats** (ventricular dysrhythmias)[20] and the actual **infarct** size (heart attack)[21] of the test animals. Notably, "this study is also the first to demonstrate that acute administration of a single dose of CBD is sufficient to reduce myocardial tissue injury irrespective of whether it is administered prior to or post coronary occlusion." Thus making CBD a novel and noteworthy possibility in further research aimed at minimizing or preventing the damage done during significant cardiac events.

Moreover, inflammation and oxidative stress are closely related processes that commonly underlie or contribute to cardiovascular disease. Studies merely using antioxidants have produced conflicting results, with some suggesting improvements, others deterioration, and still others noting no changes. To resolve the riddle of the antioxidant paradox, researchers posit that the use of antioxidants ought to be able to reduce oxidative stress and inhibit inflammations simultaneously.[22] Interestingly enough, while both THC and CBD modulate inflammatory processes and reduce oxidative stress, CBD reduces **inflammation** and **oxidative stress** without the risk of THC-induced tachycardia or anxiety and thus might be able to solve the antioxidant paradox.[23] Future studies will undoubtedly enlighten us in the near future.

Discuss with your licensed and cannabis-experienced physician if the following considerations for specific chemotypes, sub-ratios, forms, dosage, terpenes, or other plant constituents as well as their respective summaries are safe and of potential therapeutic value for you.

Chemotype-Based Considerations for Heart Disease

While both THC and CBD have been shown to possess the ability to positively reduce stress and improve mood, THC has a narrow therapeutic window, while CBD has a very wide therapeutic window.

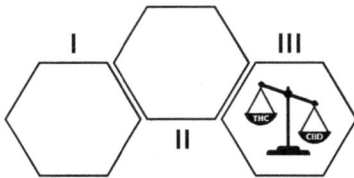

Thus, especially if you do not know your specific therapeutic amount of THC, you may avoid the potential adverse effects of tachycardia or anxiety, possibly resulting in additional cardiac stress, by choosing a CBD-rich strain (chemotype III) with no or only small amounts of THC.

Form Considerations

When considering using a tincture, make sure it contains quality tested and clearly marked amounts of cannabis constituents (e.g., 10 drops contain 5 mg CBD) so you know what you are getting; once you have discovered your optimal dose, you will be in a better position to reproduce the desired effects in a sustainable fashion.

Dose Considerations

These CBD dosage ranges are based on the human trials listed in chapter 3, section CBD.

CBD low dose: 0.4 mg to 19 mg

CBD medium dose: 20 mg to 99 mg

CBD high dose: 100 mg to 800+ mg (upper limits tested ~1,500mg)

Designing an Entourage Effect for Heart Disease

CBD is antioxidant and anti-inflammatory via mechanisms different from beta-caryophyllene, which makes beta-caryophyllene an ideal terpene to enhance both effects. Beta-caryophyllene binds with CB2 receptors of the endocannabinoid system with more potency than CBD and thus synergistically supports immune responses and is anti-inflammatory and antioxidant at once.

For additional synergistic antioxidant effects, you may want to consider: limonene, geraniol, phytol.

Linalool and geraniol may also produce subtle blood pressure-reducing effects.

Camphene may also contribute to reducing hyperlipidemia and thus may positively affect related conditions such as cardiovascular illness.

Adverse Effects and Contraindications with Pharmaceutical Drugs

Patients who have a history of atrial fibrillation, those at risk for strokes, certain heart valve diseases, or suffering from underlying blood clot conditions (e.g., pulmonary embolism, deep vein thrombosis), may be given prophylactic blood thinners such as Coumadin, which is metabolized (broken down) by a family of enzymes called cytochrome P450 2C9 (abr. CPY2C9) and pronounced "sip." CBD and THC have been found to inhibit CPY2C9 and may amplify the blood-thinning action of Coumadin (warfarin). Physicians wishing to employ cannabinoid-based therapy for these patient populations may want to exercise caution or consider reducing the blood thinner dosages. To date, very little actionable information is available. For those who wish to look at a relevant case study reported by the University of Wyoming, use the annotation for web access.[24]

Additionally, statin drugs commonly prescribed for heart disease to attempt a reduction of cholesterol are also metabolized by cytochrome P450 and may increase statin blood concentration and with it its potential therapeutic and adverse effects.

Prescribing physicians, nurse practitioners, or physician assistants are well advised to check if the pharmaceuticals used in their treatment regimen for each patient also co-administering cannabinoid-based therapies are metabolized by cytochrome P450, and if so, check blood levels and patient responses frequently and make adjustments accordingly.

Summary: Heart Disease

Step 1 Chemotype: The current trend of using an evidence-based risk versus benefit analysis leans toward the use of a chemotype III.

Step 2 THC:CBD Ratio: Most cannabis chemotype III flowers contain a low percentage of THC. An actual percentage of less than 0.3% THC will be devoid of psychotropic effects, devoid of addiction potential, and devoid of most described adverse effects (i.e. tachycardia, anxiety). Common ratios (THC:CBD) of tinctures are 1:4, 1:8, 1:15, 1:25, and higher. CBD-only tinctures are also available.

Step 3 Choose Preferred Form of Inhalation: Smoking or vaporizing the cannabis flower achieve quick and relatively short-lasting effects (about two hours). An ingestible form can produce longer-lasting effects.

Step 4 Dosing Considerations:

CBD: Low 0.4-19 mg, medium 20-99 mg, high 100-800+ mg

Step 5 Design Entourage Effect: Beta-caryophyllene, limonene, geraniol, and phytol.

Step 6: Keep careful notes on your progress and use the pain scale to measure shifts in pain or any other specific symptom (e.g., anxious). Use the worksheet to measure effects on your primary symptom before and after inhaling/ingesting. Revisit the scale at different times to better understand the subjective flow of effects in terms of duration.

Chronic (Non-Malignant) Pain

In general, pain is defined as an unpleasant physical, mental, and emotional experience that is linked to real or even only expected damage to body or psyche. In mere physical terms, pain is caused by trauma from thermal forces (heat or cold), mechanical forces (e.g., pressure, tearing, cutting), chemical forces (e.g., acids, bases, toxins), or radiation (e.g., UV light, microwave, or other ionizing radiation such as X-rays). Additionally, a significant number of people also suffer from the ill-effects of mental-emotional types of pain.

More specifically, there are a number of different types of pain (e.g., acute pain, chronic pain, central pain, peripheral pain, nociceptive pain, neuropathic pain, inflammatory pain, pathological pain). And evidence shows that no one treatment alleviates all types of pain equally well. The same is true for cannabinoid-based analgesia. For instance, evidence suggests that a cannabis chemotype II works better than a type I or type III when it comes to treating neuropathic type of pain of a chronic nature.[1]

Cannabinoids are uniquely positioned to mitigate various types of pain in all its complexities because cannabinoid receptors are widely spread across numerous pain-modulating pathways such as central and peripheral sensory neurons, brain regions modulating sensory discrimination, pain-regulatory circuitry in the brainstem, and affective states that regulate emotional responses to noxious stimuli.[2,3] In addition, the body's own endocannabinoids may provide one of the earliest therapeutic responses to pain[4] by unlocking one or more of the available pathways to realizing analgesic effects.

To cover all analgesic pathways and all types of pain is beyond the scope of this book. But we will examine one example of a particularly stubborn kind of pain that is especially resistant to standard orthodox analgesic regimens: chronic (non-malignant) pain.

The word *chronic* has its origin in the Greek *khronos*, meaning "time." Chronic pain is usually considered to last more than 12 weeks and is generally less intense than acute pain.

To better understand how speedy relief and, perhaps more important, deeper and more complete healing may be realized in cases of chronic non-malignant pain, here is a look at the evidence from these human trials.

Cannabinoid-Based Research Highlights Relevant to Chronic Pain

A randomized, double-blind, placebo-controlled crossover trial in 2004 conducted on 34 patients with chronic pain for 12 weeks demonstrated that metered sprays containing THC (2.5mg per spray) and those containing THC and CBD (2.5 mg THC plus 2.5 mg CBD per spray) were effective in relieving chronic pain. Patients were instructed to self-medicate as needed. Researchers and patients acknowledged a brief learning curve to achieve effective symptom relief while avoiding adverse effects. The effective spray usage averaged roughly eight sprays per day for THC only (total of 20 mg) and about seven sprays for combined THC and CBD (17.5 mg THC and 17.5 mg CBD). Patients were able to medicate in public without attracting unwelcome attention from others. Researchers reported that side effects were not substantially different from those seen with most other psychoactive drugs used in pain management.[5]

In 2005, a randomized, placebo-controlled, double-blind crossover trial conducted on 21 patients with chronic neuropathic pain (some with hyperalgesia and others with allodynia) utilized ajulemic acid (CT-3), a synthetic analog of a metabolite of Delta-9-THC. Like THC, ajulemic acid binds to CB1 (with high affinity, ~4× stronger than THC) and CB2 receptor sites (with moderate affinity, ~half the strength of THC).[6] Results showed that ajulemic acid was able to produce analgesic effects without occurrence of major adverse effects.[7]

A 2014 trial, using a thermal metered–dose cannabis inhaler (Syqe Inhaler Exo), conducted by a team of Israeli researchers on eight patients with chronic neuropathic pain found a 45% reduction in pain after a single dose delivering 15 mg of cannabis flower containing 19.9% dronabinol (THC), 0.1% cannabidiol (CBD), and 0.2% cannabinol (CBN). The only adverse effect reported was "tolerable lightheadedness" lasting about 15–30 minutes.[8]

In 2015 the largest trial to date, a risk versus benefit analysis, was conducted over one year on 431 patients with chronic non-malignant pain. Participants were divided into two groups, one using cannabis and the other not. Researchers examined the effectiveness and safety of cannabis with a focus on mild, moderate, and serious adverse effects, and they also monitored secondary safety measurements on lung and neurocognitive function, standard hematology, biochemistry, renal, liver, and endocrine function. Results showed that a standardized dose of dried cannabis flowers (12.5% THC; no CBD values given) up to a relatively high dose of an average of 2.5 mg a day (with a recommended top limit of 5 mg per day) for a year produced a significant reduction in pain intensity among the cannabis users over the control group. Cannabis appears in this study to have a reasonable safety profile with an increased risk of temporary mild to moderate adverse effects (such as altered state of consciousness, forgetfulness, cough, euphoria, dizziness, and nausea) in the cannabis group.[9]

In 2016 an international team from the US and Israel followed 176 patients with chronic pain and treatment with medical cannabis. After six months, median pain symptom scores decreased, pain severity decreased, pain interference scores went down, and opioid consumption was reduced by 44%.[10]

Another trial conducted on 29 patients with chronic pain in 2018 by researchers from the University at Buffalo School of Pharmacy discovered that after three months of treatment with medical cannabis improved quality of life, reduced pain and opioid use, and led to significant cost savings. "Subjects were treated with 10 mg medical cannabis capsules of THC and CBD in a 1:1 ratio, taken orally every 8 to 12 hours. If patients experienced breakthrough pain, they were prescribed a vapor pen inhaler of THC/CBD in a 20:1 ratio, 2 mg THC per 0.1 mg CBD. Patients were instructed to take 1 to 5 puffs every 15 minutes until relief was achieved and use every 4 to 6 hours as needed. Patients were prescribed a month's supply."[11]

A team of Dutch researchers from the Leiden University Medical Center conducted a randomized, placebo-controlled, four-way crossover trial in 2019 comparing various cannabis flowers with specific ratios of THC:CBD on patients with fibromyalgia. The cannabis medicine (as flower) was administered as vapor inhalation using a Volcano Medic vaporizer (Storz & Bickel GmbH & Co, Tuttlingen, Germany). Cannabis medicine used: Bedrocan (22.4 mg THC, <1 mg CBD), Bediol (13.4 mg THC, 17.8 mg CBD), Bedrolite (18.4 mg CBD, <1 mg THC), and a placebo. The complete content of the balloon was inhaled within three to seven minutes. "Bedrocan and Bediol, were analgesic in the pressure pain model." Bediol stood out by displaying a 30% decrease in pain scores when compared to the placebo.[12]

Discuss with your licensed and cannabis-experienced physician if the following considerations for specific chemotypes, sub-ratios, forms, dosage, terpenes, or other plant constituents as well as their respective summaries are safe and of potential therapeutic value for you.

Chemotype-Based Considerations for Chronic Pain

Taken together, these human trial study results suggest that observed analgesic effects in cases of non-malignant chronic pain were most likely modulated by THC, or by products containing both THC and CBD (relatively equal ratio), thus revealing the current trend of using a chemotype I or II for this kind of pain.

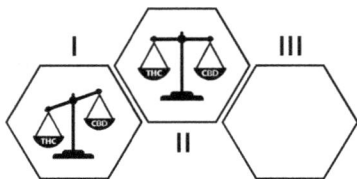

Remember, the ranges are based on limited available scientific literature to date. In fact, a number of pre-clinical studies suggest that CBD may reduce chronic pain if underlying inflammatory conditions are part of the chronic pain picture.[13,14]

Sub-Ratio

If concerns about psychotropic effects are present, consider starting with those ratios lowest in THC. For chemotype-I, consider starting carefully with an equal to or less than 5:1 ratio, and for chemotype-II, start with the lowest available level of THC (1–4%).

Form Considerations

Inhaling cannabinoids provides the fastest route to realizing analgesic effects, but their duration is relatively short (a couple of hours). When longer periods of pain relief are needed, for example to get through an eight-hour workday, an ingestible form that provides similar properties but with longer durations may be of value.

Taking the variables of absorption into account, research suggests a ratio (ingestion:inhalation) of about 1:6.[15] That is, for every 1 mg of ingested THC, about 6 mg of inhaled THC may be necessary to achieve a similar effect. Please keep in mind that this ratio is an estimate, to be considered with appropriate caution. It was derived mathematically from averages of different study results, each with ranges that had large margins. The takeaway? Use much less when ingesting.

Dosages Used in Trials

- In the first of the human trials listed here (2004), THC was used in 2.5 mg increments to a maximum of 20 mg per day.
- The second trial (2014) that mentions dosages was using an inhaler filled with a chemotype I flower (15 mg) containing ~20% THC (negligible amounts of CBD/CBN), thus delivering a maximum of 3 mg THC (15 mg × 20% THC ÷ 100 = 3 mg THC).
- The largest trial conducted (2015) using inhaled cannabis flower (12.5% THC) used up to 2.5 gm (2.5 mg = 2,500 mg) flower daily (2,500 mg x 12.5% THC ÷ 100 = ~312 mg THC). In this experiment a theoretical maximum of 312 mg THC was available. While this seems like a high average, keep in mind that much of the actual smoke is lost to the air and that the patients participating in this study were patients suffering from moderate to severe chronic pain for which orthodox analgesia was found to be inadequate or considered inappropriate and thus were likely to have a relatively high threshold for effective analgesia.
- The amount of the next trial (2018) with practical references to actual dosages was as follows: "Subjects were treated with 10mg medical cannabis capsules of THC and CBD in a 1:1 ratio, taken orally every 8 to 12 hours. If patients experienced breakthrough pain, they were prescribed a vapor pen inhaler of THC/CBD in a 20:1 ratio, 2 mg THC per 0.1 mg CBD. Patients were instructed to take 1 to 5 puffs every 15 minutes until relief was achieved and use every 4 to 6 hours as needed. Patients were prescribed a month's supply."
- The latest human trial with actionable dosage information comes to us from the Netherlands (2019). The cannabis medicine was administered as vapor inhalation. Cannabis medicine (as flower) used: Bedrocan (22.4 mg THC, <1 mg CBD), Bediol (13.4 mg THC, 17.8 mg CBD), Bedrolite (18.4 mg CBD, <1 mg THC), and a placebo. Once more, keep in mind that whenever a patient inhales, actual absorbed amounts are less due to loss of vapor to the air.

Designing an Entourage Effect for Chronic Pain

Patients suffering from chronic pain often share contributing pathologies, especially ongoing oxidative stress and inflammation. Consider strains rich in beta-caryophyllene for their additional potential to reduce oxidative stress and to fortify anti-inflammatory effects.

Adverse Effects/Contraindications with Pharmaceutical Analgesia Drugs

Any analgesic medication that produces sedative, or central nervous system depressant effects, such as Vicodin or burprenorphine used to treat opioid addiction and withdrawal sysmptoms ought to be taken with caution. Concomitant use with cannabis may enhance a number of their effects on body, mind, and emotion. For instance, cumulative sedative effects may result in altered cognitive and motor functions, making it unsafe to operate heavy machinery. However, emerging evidence also suggests a potential synergy between opioid analgesia and the use of various cannabis chemotypes for specific types of pain, thus enhancing analgesic effects while reducing the risk of opioid overdoses, adverse effects, and deaths.

CBD and THC have been found to inhibit a family of enzymes called cytochrome P450 2C9 (abr. CPY2C9) and pronounced "sip." Prescribing physicians, nurse practitioners, or physicians assistants are well advised to check if the pharmaceuticals used in their treatment regimen for each patient also co-administering cannabinoid-based therapies are metabolized by cytochrome P450, and if so, check blood levels and patient responses frequently and make adjustments accordingly.

Summary: Chronic (Non-Malignant) Pain

Step 1 Chemotype: Besides consideration of your preference for or against psychotropic effects, the current available literature suggests potential efficacy between a chemotype I, II, and to a lesser degree a chemotype III (supported by pre-clinical trials and individual case studies).

Step 2 THC:CBD Ratio: If concerns about psychotropic effects are present, consider starting with those ratios lowest in THC.

Step 3 Choose Preferred Form of Inhalation: Consider smoking or vaporizing of cannabis flower to achieve quick and relatively short-lasting effects (about two hours). If inhaling or vaping flower of unknown values, consider using 50 mg increments of actual cannabis flower to retain some level of control via the number and depth of inhalation titrated to desired effect. Remember: To obtain precision results, to avoid adverse effects, or to sustain desired effects of symptom reduction or mood improvements, you need quality-based verifiable test information about what you are consuming. Consider an ingestible form when longer duration of effects is needed.

Step 4 Optimal Amount

THC Dosage Considerations (based on dosages used in the currently available human trials)

THC micro dose: ~0.1 mg to 0.4 mg

THC low dose: ~0.5 mg to 5 mg

THC medium dose: ~6 mg to 20 mg

THC higher dose: ~21 mg to 50+ mg (for some people an upper limit is 1,000+ mg)

CBD Dosage Guidelines (based on dosages used in the currently available human trials)

CBD low dose: 0.4 mg to 19 mg

CBD medium dose: 20 mg to 99 mg

CBD high dose: 100 mg to 800+ mg (upper limits tested ~1,500 mg)

Flower Example

When choosing a chemotype-I with a relatively low THC content, if the 400 mg flower used (to fill your bong bowl) contains 7% THC, that sample contains a maximum total of about 28 mg THC (7% × 400 mg ÷ by 100% = 28 mg THC). If you are ingesting, consider starting with 5 mg THC increments (or half that if you are very sensitive or concerned about adverse effects). For those with significant concerns about psychotropic effects, consider a micro-dosing approach (for more information, go to chapter 5, step 4).

Step 5 Design Entourage Effect: Depending on your priority for an increased synergy of analgesic effects in cases of chronic pain, consider using a strain rich in beta-caryophellene.

Step 6: Keep careful notes on your progress and use the pain scale on the worksheet (p. X) to measure effects on your pain before and after inhaling. Revisit the pain scale at different times to understand the flow of analgesic effects in terms of duration.

Diabetes

Ancient physicians used the Greek word *diabetes*, which translates as "fountain," due to their observation of frequent urination in diabetic patients (the body's natural way to rid itself of excess sugar). They also noticed that their urine tasted quite sweet, as evidenced by the many ants collecting about it. *Mellitus* derives from the Greek word for "honey," hence diabetes mellitus.

Diabetes is a disease related to the pancreas, a relatively small gland behind the stomach and in front of the spine that opens into the duodenum (neck of the small intestine). Of all the glands in the endocrine system, the pancreas, along with the adrenals, sits directly in the body's center. The pancreas produces hormones such as insulin and glucagon as well as digestive enzymes that break down food into basic sugar molecules usable as food/energy by each cell of the human body.

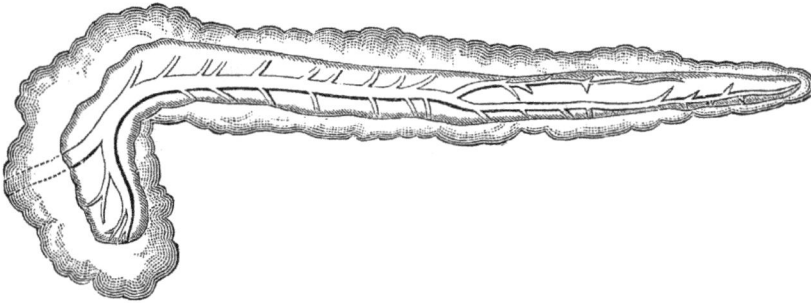

Appendix, Graphic 2: The Pancreas

Orthodox medicine lacks a cure and has not yet determined the exact origins of the illness. However, mainstream science considers a likely cause to be genetic predisposition in conjunction with some form of environmental trigger(s) or life stressor(s). One primary environmental cause is likely the emergence of industrial food production, with its decline in nutrient densities and the widespread use of pesticides and hormones and/or modern food processing and storage methods that contain endocrine disruptors. Other potential causes include certain pharmaceutical drugs, infectious pathogens, lifestyle factors (poor diet, lack of exercise, chronic stress), or an overzealous immune system that turns on itself. Higher maternal age at birth or the lack or duration of breast-feeding are also cited as possible contributing causes, particularly in cases of pediatric diabetes.

Alternatively, some researchers consider diabetes a metabolic disease involving cellular resistance to two specific hormones: insulin, which is produced by the pancreas, and leptin, which is made by adipose tissue (fat cells). Leptin signals the brain to reduce hunger. Leptin-sensitive people will feel satiated sooner and convert fat faster, while leptin-resistant individuals will eat longer and keep energy stored in fat.

Insulin tells individual cells when to turn fat or sugar into energy. Insulin-sensitive people easily convert fat and sugar into energy. In insulin-resistant people, cells do not get the message, sugar remains poorly used, and fat accumulates.

Other recent scientific discoveries suggest still another cause for the early genesis of diabetes—fatty livers. Some scientists propose that eating excessive or unhealthy sugar induces the liver to convert sugar into unhealthy types of fats, which are deposited in the liver, among other places.

Fatty livers produce insulin resistance (metabolic syndrome) in both obese and healthy people alike, increasing the risk of cancer and heart disease. Evidence suggests that avoiding sugar and eating low-glycemic foods quickly rids the body of liver/abdominal fat, and insulin resistance disappears.

Traditionally, orthodox medicine categorizes diabetes into three kinds: Type I, Type II, and gestational. Type I diabetes was once called juvenile diabetes because it occurred mostly in children or adolescents. In Type I, the pancreas stops producing the hormone insulin. Without insulin the body cannot use the food life depends on, cellular sugar. Treatment in the orthodox paradigm consists of daily insulin injections, normally administered by the patients themselves.

Type II diabetes, or adult-onset diabetes, is the most common form of the disease. In Type II the pancreas does not produce enough insulin, or the body's cells are insensitive to the presence of insulin and ignore it. This prevents the body from converting sugar into energy. Early Type II usually does not require the use of insulin; modern medical practitioners generally prescribe oral pharmaceuticals instead.

Diabetes, especially Type I or advanced Type II, is potentially life-threatening. Complications diminish one's life expectancy and quality of life. Gestational diabetes (diabetes that occurs during some pregnancies) most often self-corrects after delivery. While not very well understood, it is hypothesized that hormonal changes during pregnancy may produce a temporary resistance to insulin, thereby producing higher than normal blood sugar levels.

Early signs and symptoms of diabetes include frequent urination, sweet-smelling urine, and increased thirst and hunger. Over long periods of time, symptoms may progress to cold, pale, and clammy skin, altered levels of consciousness, unconsciousness, neuropathies, skin ulcers, peripheral vascular diseases, kidney problems, acute metabolic problems (diabetic ketoacidosis, hypoglycemia, diabetic coma, lactic acidosis), loss of vision, heart disease, stroke, infections, sepsis, and periodontal disease.

Cannabinoid-Based Research Highlights Relevant to Diabetes

Because of the well-known effects of cannabinoids on fat tissue and the glucose/insulin metabolism, it is possible that there are therapeutic applications for cannabinoid-based therapies for diabetes.[1-6] In addition, the known anti-inflammatory and immunomodulating properties of the plant have prompted a number of experiments relevant in the context of diabetes or diabetes-related conditions such as diabetic neuropathies.

While cannabis oil has been used historically in the treatment of diabetes, and while many diabetic patients claim that cannabis lowers high blood sugar levels, stabilizes mood changes, and relieves mental irritability, to date, only one double-blind, placebo-controlled human trial (13 weeks) has been conducted to examine the general effects of two cannabinoids: THCV (5 mg × 2 daily by mouth) and CBD (100 mg × 2 daily by mouth) on Type 2 diabetic patients. Of the two, THCV especially was able to improve glycemic control and pancreatic cell function. In addition, both cannabinoids at their tested dosages were well tolerated.[7]

Satiation is one of the more significant goals of an effective treatment of obesity, metabolic syndrome, and diabetes. Interestingly THC, via activation of CB1, can increase your appetite ("the munchies"), while THCV (Ki of ~75 nM) is an antagonist,[8] with about a third the strength of THC (Ki of ~25 nM) at the same receptor and thus depending on the ratio of THC:THCV, and the respective dosage produces the opposite, resulting in satiation and a consequent reduction in caloric intake. This paradoxical effect is important to consider when using flower or a full-spectrum product and trying to treat more than one symptom at the same time, such as reducing appetite and improving one's mood, or trying to lose weight while treating one's central pain.

Discuss with your licensed and cannabis-experienced physician if the following considerations for specific chemotypes, sub-ratios, forms, dosage, terpenes, or other plant constituents as well as their respective summaries are safe and of potential therapeutic value for you.

Chemotype-Based Considerations for Diabetes

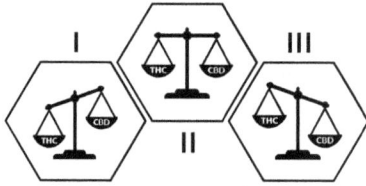

Out of the 10 studies reviewed for this section, only chemotype I and II have been used, and thus strong trend-based chemotype suggestions are lacking. However, if we consider that THC, CBD, and especially THCV may show therapeutic promise in the context of diabetes, all three chemotypes may be of relevance but especially any chemotypes rich in THCV.

Sub-Ratio

Here we are presented with important data that consider another cannabinoid (THCV) beyond that of THC and CBD. To date very limited information is available to guide our decision-making process regarding THC and CBD ratios and, perhaps more important, ratios between two cannabinoids, THC and THCV, that produce opposing effects at CB1 and with it may induce either hunger or satiation. Since satiation is a primary symptom with significant relevance to diabetes, strains low in THC and high in THCV are most commonly used.

Form Considerations

If you are opting for an ingestible form, consider choosing a product that does not contain processed or artificial sugar (e.g., soda, energy drinks, alcohol). Sugar is inflammatory. The more sugar you ingest, the more insulin and insulin-like growth factors your body makes. An increase in either can aggravate diabetes. Furthermore, consider edibles devoid of processed foods (e.g., fast foods, white bread, hormone-treated meat), which contain compounds difficult to digest and eliminate stressing the body pathways of elimination. Avoid refined gluten grains (esp. wheat, commonly treated with glyphosate, a compound in the weed killer Roundup) and dairy products, which are treated with growth hormones and thus can alter a healthy and natural hormone profile. Avoid trans fats (e.g., hydrogenated fat) which can also be highly inflammatory in nature, difficult to digest and eliminate.

Dose Considerations

If the dosages used in the abovementioned human trial are used as a guide, consider exploring dosages of THCV at 5 mg by mouth, twice daily, to reach similar results (significantly decreased fasting plasma glucose and improved pancreatic beta-cell function).

Designing an Entourage Effect for Diabetes

Beta-caryophyllene plays a regulatory role in glucose-stimulated insulin release via CB2 receptor activation and thus may be protective in Type 2 diabetes via CB2.[9]

Adverse Effects and Contraindications with Pharmaceutical Drugs and/or Other Cannabis Constituents

To date, no data exist showing the direct effects of THC or CBD on the effects of diabetic drugs. However, indirect effects, due to THC-dose-dependent CB1 activation cannabis use, may produce "the munchies," that is, increase your appetite. Eat healthy and low-glycemic index foods. Also, temporary short-term memory loss may interfere with your insulin-dosing regimen. Furthermore, THC-dose-dependent loss of coordination may make it more difficult to self-administer your subcutaneous insulin injections.

Whenever using THCV, be mindful of the fact that it is a CB1 antagonist, which may produce dysphoria.

CBD and THC have been found to inhibit a family of enzymes called cytochrome P450 (abr. CPY450) and pronounced "sip 450." Prescribing physicians, nurse practitioners, or physicians assistants are well advised to check if the pharmaceuticals used in their treatment regimen for each patient also co-administering cannabinoid-based therapies are metabolized by CyP450, and if so, check blood levels and patient responses frequently and make adjustments accordingly.

Summary: Diabetes

Step 1 Chemotype: All three chemotypes may be considered. In addition, you may especially want to consider cannabis or cannabis-containing products rich in THCV.

Step 2 THC:CBD Ratio: Since satiation is of primary concern in diabetes, whether strains are lower in THC (think munchies) or higher in THCV (anti-munchies) become important considerations.

Step 3 Choose Preferred Form of Inhalation: Consider smoking or vaporizing of the cannabis flower to achieve quick and relatively short-lasting effects (about two hours). Consider an ingestible form when longer duration of effects is needed.

Step 4 Optimal Amount:

5 mg THCV twice daily was used in the only human trial currently available .

Step 5 Design Entourage Effect: Beta-caryophyllene plays a regulatory role in glucose-stimulated insulin release via CB2 receptor activation and thus may be protective in **Type 2 diabetes** via CB2.

Step 6: Keep careful notes on your progress and use the modified pain scale on the worksheet in chapter 5 to measure effects on your primary symptom.

Anxiety

Generally speaking, anxiety is a normal reaction to the subjective experience of stress, such as in "performance anxiety." It can occur when anticipation of future events is associated in one's mind with thoughts and feelings not rooted in the present moment. While anxiety can be considered a normal part of life, acute episodes or chronic or constant anxiety can be debilitating to one's quality of life. In fact, such interference can produce very real physiological changes in the short and long term. It is estimated that almost 2 out of 10 people in the US suffer from some kind of anxiety disorder.[1]

Western medicine considers anxiety disorders mood disorders and defines five basic types: generalized anxiety disorder (GAD), obsessive-compulsive disorder (OCD), panic disorder, post-traumatic stress disorder (PTSD), and social anxiety disorder (social phobias).

Generalized Anxiety Disorder: GAD patients present with chronic worry about anticipated events constructed by the mind. Symptoms often include "feeling the other shoe is about to drop," unreasonable worry, tense and aching muscles (e.g.,, neck, shoulders), headaches, trembling, and diaphoresis (sweating).

Obsessive-Compulsive Disorder: OCD patients are characterized by compulsive, often odd behavior that can be described as personal rituals, such as filling in all the letter O's when writing, not stepping on any cracks in the sidewalk, or constantly needing to wash their hands. Continuously occupying the mind with such trivial activities provides a sense of control for those who otherwise would focus on anticipated but unwanted thoughts or feelings.

Panic Disorder: Patients with panic disorder are prone to panic attacks. Acute short-term symptoms may begin as rapid heart rate, rapid breathing (hyperventilation), and a dry mouth, sometimes progressing to an anxiety attack with restlessness, chest tightness or pain, palpitations, shortness of breath, tingling, numbness (in the hands, feet, or lips), diaphoresis (sweating), feelings of impending doom or death, and inconsolability.

Post-Traumatic Stress Disorder: PTSD develops after undergoing or witnessing a significant traumatic event. Patients frequently reexperience the trauma in their mind, either when dreaming or triggered by an external stimulus. They will often try to avoid all feelings and attempt to numb themselves against a constant dread of the return of the event. Some symptoms include "feeling on edge," flashbacks, constant fearful thoughts, sudden outbursts of anger, and an aversion to being close to anyone.

Social Anxiety Disorder: Social phobia patients are often defined by their anticipation of severely humiliating events. This type of anxiety disorder may be limited to a very narrow context, such as being extremely self-conscious about turning red (blushing) in public and then feeling judged by everybody in the room. Physical symptoms include trembling, sweating, nausea, and/or difficulty forming words or sentences.

Doctors often prescribe pharmaceuticals (e.g.,, anti-anxiety drugs, antidepressants) or psychological intervention to treat anxiety, particularly when underlying physical causes are absent. Adverse effects of pharmaceutical anti-anxiety medication range from mild to fatal.

Cannabinoid-Based Research Highlights Relevant to Anxiety

The endocannabinoid system has been implicated in the experience of anxiety in a number of scientific inquiries. Various medical traditions use cannabis extracts because of their time-proven calming and sedative effects. Today's available literature confirms that cannabinoids, those made by the human body, synthetic versions, and of course cannabis constituents, modulate mood states relevant to anxiety.

First, endocannabinoids (those made by the human body) such as anandamide especially are thought to play a central role in this effect. Anandamide, similarly to THC, binds with both of the body's own cannabinoid receptors CB1 and CB2 and produces anxiety-preventative and -mitigating effects.[2,3] For instance, research has shown that humans who are carriers of the gene FAAH 385A produce less of the enzyme that breaks down anandamide and thus have an increased amounts circulating in their blood, resulting in experiencing less anxiety.[4]

Second, synthetic cannabinoids used primarily in research have found their way into a number of "recreational uses" with often dangerous adverse effects such as severe anxiety, panic attacks, and even fatal consequences. For instance, the synthetic cannabinoid HU-210 commonly found in products advertised as herbal mixtures such as "spice" or "synthetic marijuana" binds with CB1 receptors with a magnitude 100 times that of THC.[5]

And, third, as the following paragraphs will show, a number of cannabis constituents such as THC and CBD used in the right proportions, at the proper dose, and matched to the most appropriate patient will produce an anxiety-reducing effect.

A review of the currently available studies on the effects of cannabinoids and other cannabis-based plant constituents included eight human trials, of which a majority were double-blind, placebo-controlled trials. Results from both pre-clinical and human trials show the following trends or insights.

THC and CBD, the two primary cannabinoids that tend to occur most abundantly in cannabis, have a number of distinct and physiological and psychological effects. To discover specifically what these distinctions were, researchers used a magnetic resonance imaging technique (fMRI) to examine which part of the brain was lit up by either THC or CBD. Results showed that THC and CBD did not just activate different parts (relative to placebo) in the brain's architecture but also produced opposite effects. For instance, in one task, while viewing fearful faces, subjects on either THC or CBD produced opposite objectively measurable effects in the left amygdala, fusiform, lingual gyri, the lateral prefrontal cortex, and the cerebellum, which practically translates to THC producing dose-specific anxiety and fearful reactions while CBD produced a sense of tranquility and calmness.[6]

Dosages used in the experiment were as follows: Two hours before each session, subjects had a light standardized breakfast. One hour before scanning, they were given a gelatin capsule containing either 10 mg of Δ-9-THC, 600 mg of CBD (THC-Pharm, Frankfurt, Germany), or flour (placebo). The sub-ratio was 1:60 (THC:CBD). "Pretreatments," CBD (5 mg), or a placebo were administered intravenously over 4 minutes immediately before intravenous Δ-9-THC (1.25 mg), which was also administered over 5 minutes. The sub-ratio was 1:4 (THC:CBD)

Another relevant human double-blind, placebo-controlled trial looked at a common issue central to a number of anxiety disorders. When we experience a painful event, we become hardwired to avoid any recurrences. And while this is helpful in avoiding hot plates, it is not helpful with a number of social fears. For instance, it hurts having been rejected by another person, but if the memory of that pain keeps us from ever reaching out again, we become isolated and lonely. That is quite the opposite of why we were reaching out in the first place. To unlearn that fear is a process commonly referred to by researchers as fear extinction. Now, 48 participants were conditioned with electric shocks delivered after they put their hand into a colored box. Then the participants were conditioned to unlearn their painful expectation every time after they put their hand into that box. Results showed that 32 mg of CBD significantly supported extinction learning, thus suggesting a novel treatment possibility for anxiety disorders.[7]

In another double-blind, placebo-controlled human trial using neuroimaging techniques on 10 patients diagnosed with social anxiety disorders (SAD), researchers discovered that an oral dose of 400 mg CBD significantly decreased the subjective experience of anxiety. Researchers pointed out that the imaging results suggest that CBD reduces anxiety in SAD due to its effects on activity in limbic and paralimbic brain areas.[8]

Researchers from New Zealand performed gentle osteopathic manipulations on 37 healthy people and measured their endocannabinoid levels in response to it. Results showed a significant increase in anandamide levels and corresponding expansive emotional experiences, leading the scientists to speculate that their osteopathic treatments were modulated via the ECS and thus play a role in mood improvements.[9]

Since the fear of public speaking is so prevalent, it is well suited to be used in determining what might help. Fifty-seven healthy males were given oral doses of CBD at either 150 mg, 300 mg, 600 mg, or a placebo in a double-blind procedure before having to give a speech. Results showed that compared to the placebo, pretreatment with 300 mg of CBD significantly reduced anxiety during the speech.[10]

The trial echoes results from an earlier double-blind placebo-controlled trial in which 24 patients diagnosed with social anxiety disorders were tasked with public speaking. Half of them were given 600 mg of CBD; the other half were given a placebo one hour before the speech was to commence. Results showed that pre-treatment with CBD significantly reduced anxiety, cognitive discomfort, and cognitive impairment.[11]

And the latest double-blind, placebo-controlled human trial comes from Italy where researchers studied the effects of essential oil of cannabis on the brain and nervous system. Results showed that inhalation of primarily THC (<0.2%w/v) as well as those terpenes that occurred in their highest concentrations, namely myrcene (22.9%) and beta-caryophyllene (18.7%), decreased diastolic blood pressure, increased heart rate, and significantly increased skin temperature. The subjects described themselves as more energetic, relaxed, and calm, suggesting that even just inhaling essential oil of cannabis produces changes in the autonomic nervous system that can be harnessed to improve moods and mitigate anxiety.[12]

Discuss with your licensed and cannabis-experienced physician if the following considerations for specific chemotypes, sub-ratios, forms, dosage, terpenes, or other plant constituents as well as their respective summaries are safe and of potential therapeutic value for you.

Chemotype-Based Considerations for Anxiety

The amount of THC, the primary psychoactive cannabinoid of cannabis, is the source of major concern and focus in determining a safe and optimal dose. As such, a chemotype I with more THC than CBD has indeed a very narrow therapeutic window where the line between a therapeutic and an adverse effect is razor thin. Even a chemotype II with a balanced THC and CBD ratio and a medium-size therapeutic window may still be too much for some sensitive patients. A chemotype III (with very low to trace amounts of THC %) may provide the safest option.

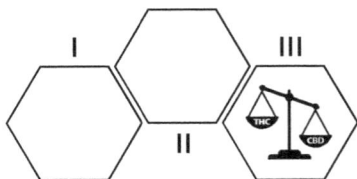

Sub-Ratio

Patients who are very concerned about cognitive changes tend to use a CBD-only product or start with a 1:25 (THC:CBD) ratio flower or tincture.

Form Considerations

While cannabis-containing edibles tend to provide longer-lasting effects, caution is advised. Unless you or your treating physician knows the optimal dose of the primary psychoactive THC, proceed slowly and in smaller increments, waiting up to two hours in between adding another dose.

Dosage Considerations

The researchers who conducted the anxiety-related human trials listed previously utilized oral forms of either a cannabis chemotype III with a THC:CBD ratio of 1:60 (dosages of 10 mg THC:600 mg CBD) or used oral forms of CBD only (dosages ranging between 32 mg to 600 mg).

Designing an Entourage Effect for Anxiety

Inhaling essential oil of cannabis containing relatively high concentrations of **myrcene** (22.9%) produced physiological changes such as decreased diastolic blood pressure, increased heart rate, and increased skin temperature; the human subject felt more energetic, relaxed, and calm, suggesting that inhaling these scent molecules may contribute to changes in the autonomic nervous system that can be harnessed to improve moods and mitigate anxiety.[13]

The same study also pointed to relatively high concentrations of **beta-caryophyllene** (18.7%) as a potential contributing plant constituent to produce the same effects noted in the paragraph above.[14] Additionally, pre-clinical trials conducted on mice are suggesting that beta-caryophyllene may have CB2-initiated anxiolytic and anti-depressant effects. If these effects are confirmed in humans, this terpene may be relevant in cases of mood disorders.[15]

A study from Japan conducted on mice registered a significant anxiolytic effect that was triggered via olfactory pathways evoked by the odor of **linalool**.[16] Another animal study, where mice that inhaled linalool from ambient air for one hour showed a significant reduction in motility, suggesting a significant sedative effect.[17]

Phytol has been shown to be able to inhibit the enzyme that breaks down GABA (a natural neurotransmitter responsible for calming effects); it may increase GABA's bioavailability and produce mildly relaxing effects.[18]

As early as 1972, French researchers described the sedative and antispasmodic properties of **nerolidol**.[19] One study conducted on mice discovered that inhaled essential oil of *Microtoena patchoulii* leaves produced a sedative effect, and that **terpinolene**, its active ingredient, was essential to produce sedation.[20]

Numerous pre-clinical and clinical models demonstrated that **d-limonene** could produce anti-stress effects via modulation of parasympathetic and central neurotransmitter functions.[21] Findings from another experiment conducted on mice suggest that a mixture of cis and trans of (+)-limonene epoxide reduced anxiety without any observable adverse effects.[22] A laboratory study discovered that d-limonene directly binds to the adenosine receptor, which may induce sedative effects.[23]

Adverse Effects and Contraindications with Pharmaceutical Drugs

Cannabis or cannabis-containing products have the potential to increase or decrease the potency and therefore the effectiveness and adverse effects potential of anti-anxiety pharmaceuticals (as well as other recreational drugs such as alcohol). For example, a patient on valium may feel overly sedated or drowsy when additionally using cannabis. Or, for instance, if Prozac was prescribed to reduce anxiety and elevate the patient's mood, co-treatment with cannabinoids may increase the bioavailability of the drug and thus increase the risk of extrapyramidal reactions (involuntary movements).

CBD and THC have been found to inhibit a family of enzymes (to date over 50 members have been identified)[24] called cytochrome P450 (CYP450) and pronounced "sip 450," which metabolizes (breaks down) a great number of pharmaceutical drugs, among them various anxiolytics. Prescribing physicians, nurse practitioners, or physician assistants are well advised to check if the anxiolytic pharmaceuticals used in their treatment regimen for each patient are metabolized by a member of the family cytochrome P450, and if so, check blood levels and patient responses frequently and make dosage adjustments accordingly.

Warning:

It is important to remember, especially in the context of anxiety, that while too small amounts of THC can be sub-optimal, too much can actually increase anxious feelings and lead to unpleasant adverse effects that can last for hours.

Summary: Anxiety

Step 1 Chemotype: Based on the current evidence-based trends, a chemotype III is the potentially safest anxiolytic option. However, if you know your therapeutic window (i.e. your optimal THC dose), you may find the sedative effects of a THC-rich type a good fit to meet your goal of relaxation and/or mood elevation.

Step 2 THC:CBD Ratio: People very concerned about psychoactivity often choose a CBD only product or start with a 1:25 (THC:CBD) ratio flower or tincture.

Step 3 Choose Preferred Form of Inhalation: People looking to achieve quick and relatively short-lasting effects (about two hours) tend to consider inhalation, while those who desire longer-lasting effects tend to consider an ingestible form.

Step 4 Optimal Amount: Depends largely on each person's needs and preferences. Amounts of oral CBD used in human trials ranged between 32 and 600 mg.

Step 5 Design Entourage Effect: Consider using myrcene, beta-caryophyllene, linalool, phytol, nerolidol, terpinolene, and/or limonene.

Step 6: Keep careful notes on your progress. Consider using a modified pain scale on the worksheet in chapter 5 to measure effects on your primary anxiety symptom (e.g., tension, constant presence of fearful thoughts even when no danger is present) before and after use. Consider revisiting the pain/symptom scale at different times to better understand the flow of effects.

Neurological Conditions

Orthodox medicine considers multiple sclerosis a chronic, inflammatory, and degenerative neurological illness with no cure and no exact cause. In fact, MS is one of the most common neurological diseases.

The meaning of the term *multiple sclerosis* provides a clue about the general picture of this illness. It derives from the Latin words *multi* and *plus*, which together translate into "manifold," and the Greek word *sclerosi,* which translates as "hardness." Place the words together, and we get "many folded hardness." Apply "manifold hardness" to the brain and spinal cord, and we have a description of MS.

Multiple sclerosis is characterized by the breakdown of some of the thin sheets that cover the brain and spinal cord. These fat-based myelin sheets normally provide insulation and protection, but when lesions occur, nerve impulses misfire across the broken insulation, causing a variety of debilitating symptoms.

Scientists speculate that MS might be an autoimmune disease that inadvertently prompts adhesion molecules, which summon immune cells to fight an inflammation and thus contribute to the destruction of myelin sheets.[1] Other likely culprits include a combination of genetic factors, infections, and environmental influences such as decreased sun exposure and subsequent insufficient vitamin D.

Some pharmaceutical medications exist to manage symptoms as they develop or worsen, but they come with the price of potentially significant adverse effects. MS may initially appear as acute attacks only, or it may progress to its chronic degenerative form in which symptoms accumulate and gradually worsen. Some of the most common symptoms include muscular spasms affecting the eyes, bladder, and bowels, numbness, increased weakness, ataxia, slurred speech, and/or acute and chronic pain.

Due to limited success within the allopathic model, many MS patients have turned to other healing methodologies in hopes of finding help and relief.

Cannabinoid-Based Research Highlights Relevant to Multiple Sclerosis

The earliest study listed in the National Library of Medicine took place in 1981, when researchers found motivation in anecdotal accounts of MS patients who reported that inhaling cannabis gave relief from spasticity. This, combined with the scientific discovery that THC is able to inhibit spasms in animal studies, opened the door to a multitude of scientific inquiries.

While initial studies merely focused on observing the effect of cannabis on the most common symptoms of MS, later studies worked to discover the mechanisms underlying the observed therapeutic effects.

The scientific community began to build on the data accumulated. The therapeutic frame of cannabis in the context of MS became clearer and more defined. A reduction in spasticity was recognized as only the first of many benefits. Cannabis use also reduced pain, depression, anxiety, paresthesia (e.g., abnormal sensation such as tingling), neuropathic pain, sleep disturbances, urge incontinence, and nocturnal polyuria.

At the time of this writing, over 70 studies with relevance to multiple sclerosis were available. And while these individual studies differ in approach, scope, and methodology, the results share a common denominator. For patients suffering from MS, cannabis may be a potent ally in alleviating the disease's primary and secondary symptoms as well as slowing the progression of the illness itself and thus providing an improvement to long-term survival and quality of life.

Warning: Most clinical studies report adverse effects of cannabis use in addition to benefits dependent on dose and form. Adverse effects include reduced balance and posture, nausea, and dizziness; and at high dosages, negative psychological symptoms such as anxiety. These studies indicated that when used within the proper subjective therapeutic dose, cannabinoids' potential adverse effects are usually well tolerated and negligible, especially when compared to the beneficial effects.

Discuss with your licensed and cannabis-experienced physician if the following considerations for specific chemotypes, sub-ratios, forms, dosage, terpenes, or other plant constituents as well as their respective summaries are safe and of potential therapeutic value for you.

Chemotype-Based Specific Considerations for Multiple Sclerosis

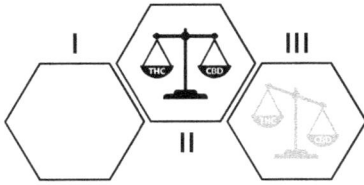

Of the 70-plus studies reviewed, 15 highlighted a chemotype I, 32 a chemotype II (mostly in the form of Sativex), and 2 a chemotype III. The emerging trend here is the use of a chemotype II. Sativex is a whole-plant extract that contains relatively equal amounts of THC and CBD.

Sub-Ratio

Sativex has a ratio of ~1:1 (THC:CBD).

Form Considerations

Inhalation will deliver a therapeutic effect at a very rapid rate but usually tapers down in a couple of hours. Ingestion will produce longer-lasting effects. Start slow and use caution if you are new to using an ingestible product.

Dosage Considerations

One iteration of Sativex delivers ~2.7 mg THC : ~2.5 mg of CBD. The manufacturer's leaflet for Sativex suggests starting with a gradual increase (trial) over two weeks (leave at least 15 minutes between sprays):

	Number of Sprays				
Days	AM (morning)	PM (evening)	Total Sprays per Day	Total Daily THC amount	Total Daily CBD amount
1			1	2.7 mg	2.5 mg
2			1	2.7 mg	2.5 mg
3			2	5.4 mg	5.0 mg
4			2	5.4 mg	5.0 mg
5			3	8.1 mg	7.5 mg
6			4	10.8 mg	10.0 mg
7	1	4	5	13.5 mg	12.5 mg
8	2	4	6	16.2 mg	15.0 mg
9	2	5	7	18.9 mg	17.5 mg
10	3	5	8	21.6 mg	20.0 mg
11	3	6	9	24.3 mg	22.5 mg
12	4	6	10	27.0 mg	25.0 mg
13	4	7	11	29.7 mg	27.5 mg
14	5	7	12	32.4 mg	30.0 mg

For warnings, precautions, and more specific product information, consider going online and reading the Sativex leaflet from the producer (Bayer/GW Pharmaceuticals).

Alternatively, a chemotype II flower can be found at some dispensaries containing various percentages of THC:CBD (current maximum of up to 12% THC/CBD).

Designing an Entourage Effect for Multiple Sclerosis

Consider beta-caryophyllene, limonene, linalool, and pinene.

Interactions with Pharmaceutical Drugs

Since there is no cure for MS within the orthodox system of medicine, treatment is focused of on managing symptoms with pharmaceuticals such as corticosteroids aimed at reducing inflammation and to modulating an overly reactive immune response common in MS. Gluccocorticoids are metabolized by members of the cytochrome P450 enzyme family, and thus co-administration of cannabinoids (THC/CBD) may require dosage adjustments.

Summary: Multiple Sclerosis

Step 1 Chemotype: The current evidence-based trends indicate a therapeutic potential for the use of a chemotype II followed by a chemotype I. Currently available chemotype II strains top out at around 12% THC and CBD.

Step 2 THC:CBD Ratio: Sativex is ~1:1.

Step 3 Choose Preferred Form of Inhalation: Consider smoking or vaporizing of cannabis flower to achieve quick and relatively short-lasting effects (about two hours). Consider an ingestible form when a longer duration of effects is needed.

Step 4 Optimal Amount: If you are ingesting, consider the sativa regimen starting with 2.5 mg THC increments. Or, for those with significant concerns about psychotropic effects, consider a micro-dosing approach (go to chapter 5, step 4)

Step 5 Design Entourage Effect: Beta-caryophyllene, limonene, linalool, and pinene.

Step 6: Keep careful notes on your progress and use a modified pain scale on the worksheet in chapter 5 to measure effects on your primary symptom before and after inhaling. Revisit the scale at different times to understand the flow of effects in terms of duration.

Insomnia

Insomnia is difficulty with or inability to fall or stay asleep. Also, poor quality of sleep resulting in waking up tired or exhausted is a common symptom of people suffering from insomnia.

It is estimated that in the United States alone, a third of all adults suffer from insomnia at some point in their lives. Women are suffering from insomnia at twice the rate than men do, and over half of all seniors over 60 suffer from this condition.

Western medicine divides insomnia into primary and secondary insomnia. In primary insomnia, insomnia is the sole problem. Secondary insomnia is produced by other causes.

Insomnia is further divided into acute and chronic insomnia. Acute insomnia is considered temporary, short-term, and self-correcting, resulting from new work hours, jet lag, and significant time-zone differences. Chronic insomnia has an ongoing and underlying problem such as numerous diseases, pharmaceutical or recreational drug use, or pain.

Cannabinoid-Based Research Highlights Relevant to Insomnia

The results of a review of over 25 sleep-wake cycle, insomnia, and cannabinoid studies provide the current evidence to help us make more informed decisions in relation to insomnia. The trends from the key study results indicate that effects of using individual cannabinoids are dose-dependent. Too much or too little of a single cannabinoid can have unwanted effects in general and in time specifically, as in biphasic effects where initially you may find relief but a couple of hours later you find your symptoms are getting worse. The results also show that two of the prime cannabinoids, THC and CBD, can create a synergy. It seems that both taken together, or even better as a full-spectrum product, balance each other so as to maximize the most desired effects: ease of falling asleep, deep sleep, and no residual drowsiness upon morning rising. Here are some of the more specific findings:

- THC promotes deep sleep.[1,2,4,8,9] Anandamide promotes deep sleep.[7]
- THC reduces the time it takes to fall asleep.[12,13]
- However, THC may also cause morning drowsiness.[3]
- In contrast, CBD was found to promotes wakefulness[8,10] at low doses while supporting deeper sleep at higher dosages.[5,6] CBD also improves the quality of sleep.[11]
- CBN, a product of THC degradation over periods of time, may promote sleep.[15,16]
- THC can induce anxiety and panic, especially when used in higher dosages and in susceptible patients.[14]

Discuss with your licensed and cannabis-experienced physician if the following considerations for specific chemotypes, sub-ratios, forms, dosage, terpenes, or other plant constituents as well as their respective summaries are safe and of potential therapeutic value for you.

Chemotype-Based Considerations for Insomnia

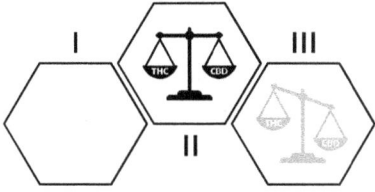

Creating the best possible sleep synergy by discovering the best chemotype for insomnia depends on your particular subjective needs, which can differ significantly from one patient to another. However, with a bit of attention to detail, it may be easier than you think. We know that THC promotes sleep, reduces the time it takes to fall asleep, but can also cause morning drowsiness; and at higher dosages, especially for susceptible patients, it can cause panic and anxiety, thus making insomnia worse. CBD at higher dosages also supports deep sleep but at lower dosages induces wakefulness and thus balances out the sedating qualities of THC.

Thus many people opt for starting with a chemotype II with a relatively equal THC:CBD ratio. If you are very concerned about mind- or mood-altering adverse effects, consider using a chemotype III with a low ratio of THC to CBD. If you are only moderately concerned about said adverse effects, you may still want to consider a chemotype II, but consider starting with one containing a low level of THC. Either way, allow yourself to be guided by tracking your experiences and sensations in your journal or on the worksheet in chapter 5.

However, many people who have had good experiences using the old system of selecting cannabis strains have most likely used indicas and indica hybrids as a sleeping aid. The insomnia go-to strains such as Blackberry Kush, Purple Urcle, Godfather Og, or Blackberry Hashplant, for example, tend to have relatively high concentrations of THC and low concentrations of CBD. If these strains worked for you, you will get good results with a chemotype I. If you have no morning drowsiness or anxiety problems with the use of this type, it may make sense to stay with it, including the amount and form that worked for you. However, if you need more alertness in the morning, consider stepping down ratios (step 2) by one level. For instance, if you used a THC:CBD ratio of 100:1, step down to a strain/product with a ratio of about 50:1. Write down your experiences and sensations in your journal and make adjustments accordingly. You can use the guiding questions as a framework to navigate toward your optimal chemotype.

Sub-Ratio

Sub-ratios can vary depending on your subjective experience and can range along the full spectrum of common THC:CBD ratios (e.g., 100:1, 50:1, 20:1, 5:1, 1:1, 1:4, 1:8, 1:15, 1:25).

Additional Considerations

Cannabinol (CBN)

Most CBN found in cannabis plant material is due to chemical changes that naturally occur during the drying process. Over time, THC begins to degrade, and some of its chemical composition transforms into CBN. Cannabinol may promote sedation. CBN is not generally considered a mind-altering cannabinoid or at best is a very mild psychoactive compound. However, CBN has not been studied in the context of insomnia, and thus specific guidance is wanting.

Anandamide

Research posits that mind-body medical techniques such as osteopathic treatments that induce mood-altering effects, relaxation, and analgesia may do so by significantly increasing endocannabinoid levels such as that of anandamide, the body's own version of THC.[17] In addition, since stress is almost always a known or suspected factor with insomnia, consider an appropriate exercise for you—for example, strenuous exercise significantly raises anandamide levels naturally.[18,19] Mitigating and balancing the effects of stress through mindfulness techniques such as meditation, tai chi, or yoga, might also help with insomnia symptoms by modulating the endocannabinoid system, increasing anandamide levels naturally.[20]

Form Considerations

Inhaling brings relatively quick results that tend to fade after a couple of hours. So for those people who just need a little help falling asleep but who have no problem staying asleep, this might be a good option to consider. However, due to the intricate nature of insomnia symptoms (and possible underlying conditions) and the very real possibility of biphasic effects, using cannabis flowers with precision and in a sustainable fashion is more complex than using a longer-lasting tincture with clearly measurable and already defined portions and proportions. To begin, it helps to know the total amount of THC and CBD in your loading dose.

For instance, to determine the total amount of THC and CBD in your chemotype II flower, assume your loading dose of cannabis flower is 200 mg for a vaporizer, bong, or a joint. If the concentration was 12% for instance, then your total THC content (and in this case for CBD, since it is a 1:1 ratio) would approximate 24 mg (12% × 200 mg = 2,400 ÷ 100% = 24 mg).

In contrast, and for those people who tend to wake up frequently and thus need longer-lasting results, you may want to consider an ingestible form. Cannabis-infused tinctures, oils, and sublingual sprays can work very well, and according to compliance requirements in more and more states the ratio, concentration, and dosage amounts are published on the container. Thus, once you have found your specific dosage, these products may better provide you with effective and sustainable relief.

Taking the variables of absorption into account (such as typical loss rates during digestion, as well as an estimation of the impact of 11-OH-THC), research suggests a ratio (ingestion:inhalation) of about 1:6.[21] That is, for every 1 mg of ingested THC, about 6 mg of inhaled THC may be necessary to achieve a similar effect. The reader is asked to keep in mind that this ratio is an estimate, to be considered with appropriate caution. It was derived mathematically from averages of different study results, each with ranges that had large margins. As with ingesting cannabis-infused tincture or oils (or products that contain them such as cookies), consider starting with about 2.5 mg to 5 mg increments of THC. Wait for one hour (two hours if your stomach is full). Note any changes in your journal. Repeat until you achieve the desired effect. Note in your journal exactly how many drops you took. Stop and regroup if you notice unwanted effects.

Designing an Entourage Effect for Insomnia

Myrcene has sedative effects on its own. The terpene alone is not as strong as the cannabinoids, but working together, it can increase the calming and relaxing sensations of the other cannabinoids.

Guiding Questions

No matter what form you choose, but especially if you are choosing cannabis flowers as your preferred form, consider writing down answers to these questions in your journal or the worksheet so it will be easier for you to keep track of your progress. Here is a cannabis chemotype II as an example.

- How long did it take you to fall asleep with this chemotype II?

  Quick, medium, long? If it took a long time to fall asleep, you may want to consider an increase in THC.

- How long did I sleep? Hours (shooting for eight)?

  If fewer, you may want to consider an increase in THC.

- Did you sleep through and soundly? Yes or no?

  If not, consider an increase in THC.

- Did you wake up refreshed? Yes or no?

  If not, consider increases in CBD to reduce morning drowsiness. You may find it helpful to switch to a chemotype III with more CBD and less THC than what you have been using so far.

Also, remember to go slow. Consider adjusting the ratios or concentration in accordance with your experiences and make adjustments one step (one level) at a time. Listen to your sensation, feeling, and symptoms. It may take a few days of journaling to find the ratio and concentration that is right for you. But with a little patience and following your own subjective instructions as you wrote them down, your journal will give you the feedback you need.

Adverse Effects and Contraindications with Pharmaceutical Drugs

Remember that both THC and CBD can have a dose-dependent biphasic effect. For instance, a low dose of CBD can produce wakefulness, while higher dosages tend to produce deeper sleep. Taking too high a dose, the wrong ratio, or the less optimal chemotype may lead in some cases to an increase in insomnia, morning drowsiness, anxiety, and even panic attacks.

Additionally, after using cannabis-based preparations for a few weeks, rebound insomnia may occur if you stop using them or if underlying causes are not resolved.

THC- or THC-abundant products especially can increase the effects of alcohol and pharmaceutical sleeping aids commonly prescribed. Consult with your prescribing physician and exercise caution when you are taking benzodiazepines (e.g., Valium, Halcion, or Xanax), non-benzodiazepine sedative-hypnotic drugs (e.g., Ambien, or Restoril), or antidepressant and sleep aid combination drugs such as selective serotonin reuptake inhibitors (SSRIs) (e.g., Elavil or Doxepin) and even over-the-counter antihistamines such as Benadryl or Tavist.

Summary: Insomnia

Step 1 Chemotype: Consider starting with a chemotype II (especially if concerned about feeling anxious). However, depending on your specific needs and symptom presentation, all chemotypes may apply.

Step 2 (THC:CBD Ratio): Sub-ratios can vary depending on your subjective experience and can range along the full spectrum of common THC:CBD ratios (e.g., 100:1, 50:1, 20:1, 5:1, 1:1, 1:4, 1:8, 1:15, 1:25).

Step 3 (Choose Preferred Form of Inhalation): For rapid, shorter-term relief, consider inhalation. For longer lasting-results consider an ingestible.

Step 4 (Optimal Amount): If inhaling cannabis flower of unknown and untested concentration, consider 50 mg increments (of cannabis flower) and titrate to desired effects.

Remember, to obtain precision results, to avoid adverse effects, or to sustain desired effects of symptom reduction or mood improvements, you need quality-based verifiable test information about what you are consuming. Use caution to avoid possible unwanted biphasic effects

Step 5 (Design Entourage Effect): If ingesting, consider starting with 2.5 mg THC to no more than 5 mg THC. Be patient and progress slowly when adding more (wait one to two hours in between) to avoid potential adverse effects.

Step 6: Consider strains rich in myrcene when additional sedative effects are needed.

Gastrointestinal Conditions—Colitis

Colitis, also commonly referred to as inflammatory bowel disease (IBD), simply means inflammation of the large intestine. There are a number of types of colitis which are identified by their underlying causes, such as autoimmune colitis, ischemic colitis, allergic colitis, microscopic colitis (i.e., collagenous colitis, lymphocytic colitis), ulcerative colitis, chemotherapy-induced colitis, infectious colitis (e.g., salmonella), or idiopathic colitis (of unknown origins).

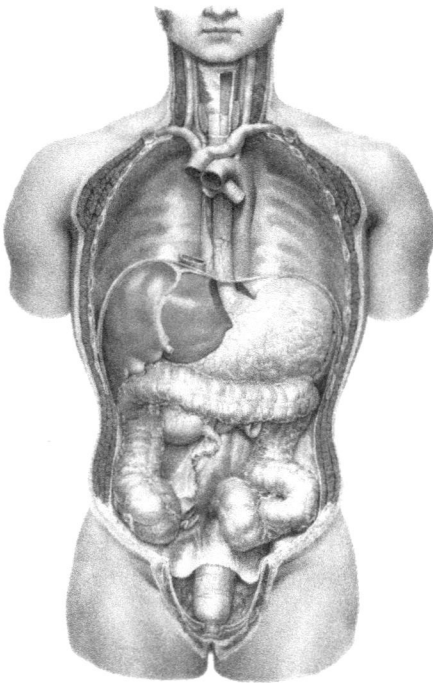

Symptoms commonly include abdominal pain or cramps, diarrhea (may include blood, mucus, or pus), constipation (e.g., difficult BM despite urge), generalized weakness, weight loss, anemia, inflammation, nausea, stress, or lack of appetite.

To date, there is no cure within the model of modern medicine, and treatment is limited to supportive therapies such as hydration via IV fluids, pharmaceutical intervention (e.g., steroids to depress the immune system), iron supplements to balance anemia, or surgery. Some progress has been reported with certain dietary regimes and microbiome supportive supplements. Colitis is different from irritable bowel syndrome (IBS), a chronic and typically non-inflammatory syndrome of the large intestine posited to be of psychosomatic origins.

Appendix, Graphic 3

Orthodox diagnoses are performed by elimination. Doctors run a variety of tests to rule out diseases with similar symptoms. These may include colonoscopies, screening for parasites (blood or stool tests), testing for lactose intolerance (hydrogen breath test) or the presence of infections (stool examinations), as well as tests for celiac disease (blood test screening for antibodies). If none of these diseases are responsible for the patient's symptoms, practitioners may follow one of several possible established diagnostic algorithms (a list of questions related to the patient's symptoms).

Physicians typically manage the disease with dietary modifications, pharmaceutical medications, and referrals to psychotherapy. Alternatively, Canadian researchers conducted a meta-analysis of all randomized controlled trials published on Medline, Embase, and the Cochrane register up through April 2008. They reported that fiber, antispasmodics, and peppermint oil exhibited greater effectiveness than a placebo in the treatment of colitis.[1]

Cannabinoid-Based Research Highlights Relevant to Colitis

Case reports from cannabis-using colitis patients suggest that cannabis may be effective in managing some symptoms, especially nausea, diarrhea, stress, cramps, weight loss, and lack of appetite.[2-4] Pre-clinical trials suggest therapeutic potential of cannabinoids on forms of colitis.[5-7]

Three observational studies conducted on about 300 adult subjects suggest that cannabis has subjective therapeutic benefits.[8-10] Additionally, Italian researchers conducted a meta-analysis/review of the available pre-clinical studies related to cannabinoids and the gut. The authors wrote: "Anatomical, physiological, and pharmacological studies have shown that the endocannabinoid system is widely distributed throughout the gut, with regional variation and organ-specific actions. It is involved in the regulation of food intake, nausea and emesis, gastric secretion and gastroprotection, GI motility, ion transport, visceral sensation, intestinal inflammation and cell proliferation in the gut."[11]

A placebo-controlled clinical trial conducted on 21 patients with significant manifestations of colitis who did not respond to the usual pharmaceutical regimen consisting of therapy with steroids, immunomodulators, or anti-tumor necrosis factor-α agents. Results showed that inhaling from a cannabis cigarette (containing 115 mg of THC twice daily) produced complete remission in 5 of 11 patients in the cannabis group and 1 of 10 in the placebo group (remember when smoking the actual THC amount absorbed is significantly less than the total amount of THC present in the cigarette). Three patients in the cannabis group were weaned from steroid dependency. Additionally, patients receiving cannabis reported improved appetite and sleep, with no significant side effects.[12]

Discuss with your licensed and cannabis-experienced physician if the following considerations for specific chemotypes, sub-ratios, forms, dosage, terpenes, or other plant constituents as well as their respective summaries are safe and of potential therapeutic value for you.

Chemotype-Based Considerations for Colitis

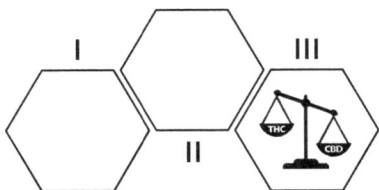

Of the 27 studies reviewed for this section, 2 used a chemotype I, 1 a chemotype II, and 6 used a chemotype III, giving the lead of the current trends to chemotype III, or CBD-abundant products.

Sub-Ratio

Many patients have been reporting good results using an ingestible product (e.g., tincture, capsules) with a ratio of THC:CBD of 1:20 for daily support and then, depending on the results increase or decrease the THC. Additionally, you may want to consider using a cannabis chemotyope III with small amounts of THC (e.g., 1:15) to address flare-ups.

Form Considerations

If you are choosing an ingestible product, consider choosing one that is devoid of materials that could work against producing a therapeutic effect, such as refined sugar or artificial sweeteners, glyphosate-treated foods (esp. wheat), dairy (esp. with diarrhea), unhealthy fats (esp. trans fats), caffeine, and carbonated drinks.

Inhalation can yield rapid results and allows you to titrate to your specific effect. To be as safe as possible, stop inhaling when your desired effect is achieved to avoid adverse effects. Inhalation is also sometimes used by colitis patients to address shorter-term flare-ups. Alternatively, you may also use sublingual products (e.g., sprays, strips) to reach similar results.

Dosage Consideration

Forms of colitis tend to express themselves with great variations and fluctuations, and thus there is no silver bullet that fits all approaches to dosing. Furthermore, as the internal environment (esp. your microbiome, pro- and anti-inflammatory responses, and immune system) responds and shifts to the introduction of cannabinoids, what worked last week may be in need of adjusting this time around. It may take some time and experience to get good at dialing in the appropriate dose, ratio, and other plant constituents. However, this attention to detail recorded in your session log or journal is well worth the effort to regain balance and resilience when it comes to digesting, absorbing, and eliminating food and drink as well as emotional materials.

Below are commonly used CBD dosage ranges (to test your sensitivity, start low and build up slowly):

- CBD low dose: 0.4 mg to 19 mg
- CBD medium dose: 20 mg to 99 mg
- CBD high dose: 100 mg to 800+ mg

Designing an Entourage Effect for Colitis

At least one of the studies reviewed suggested that CBD in collaboration with other cannabis constituents included but not limited to CBC, CBG, and THC were superior to CBD alone.[13]

Beta-caryophyllene is a CB2 agonist and produces anti-inflammatory effects.[14] Pre-clinical trials determined that beta-caryophyllene may provide a novel treatment possibility in cases of colitis by initiating therapeutic actions via CB2 and PPARγ receptor pathways.[15] beta-caryophyllene was found to be effective against Candida albicans, a common fungal infection in an unhealthy gut environment.[16]

Limonene has been reported to dissolve cholesterol-based gallbladder stones, reduce stomach acidity, and to normalize peristalsis (gastrointestinal problems) for patients suffering from heartburn and GERD.[17]

Research also discovered that beta-myrcene may play a role in preventing peptic ulcer diseases.[18]

Peppermint oil has been found to be helpful for some patients with IBS. Peppermint oil is a compilation of several individual terpenes, including limonene, pinene, and cineol, which also are found in cannabis.

Adverse Effects and Contraindications with Pharmaceutical Drugs

The understanding of the interaction of cannabinoids and pharmaceutical drugs is in its infancy. However, early studies indicate that cannabinoids such as CBD or THC interact in a variety of ways (e.g., inhibit, block, enhance) with a family of enzymes (Cytochrome 450) that breaks down a majority of all pharmaceutical drugs on the market. Thus high dosages of an isolated cannabinoid may make a pharmaceutical drug less effective and/or more toxic.

If a patient with chronic diarrhea symptoms utilizes anti-diarrheal medication containing opioids, caution is advised. THC may increase opioid-based sedative effects. Furthermore, THC and CBD may produce synergistic analgesic effects allowing more effective pain control at lower than normal opioid dosages.

Serotonin is primarily produced in the gut. CBD has been shown to facilitate serotonin receptor mediated neurotransmission, and may reduce the effects of serotonin antagonist, another common prescription in cases of stress-induced diarrhea. However, since serotonin enhances motility (bowel movements), this CBD serotonin enhancement may serve those suffering from constipation.

Since diet-induced obesity produces changes in the healthy gut microbiome, probiotics may be therapeutic in cases of colitis. Research has shown that both were prevented by ongoing administration of THC, suggesting a synergistic action between microbiome and THC in maintaining gut health in general.[19]

Summary: Colitis

Step 1 Chemotype: The current trends highlight a chemotype III.

Step 2 THC:CBD Ratio: Consider starting with a 1:20 (THC:CBD) ratio.

Step 3 Choose Preferred Form of Inhalation: Consider smoking or vaporizing of cannabis flower to achieve quick and relatively short-lasting effects (about two hours). Consider an ingestible form when a longer duration of effects is needed.

Step 4 Optimal Amount: Consider starting with a low dose range (0.4 mg to 19 mg) and slowly increasing CBD amounts to a CBD medium dose range (20 mg to 99 mg) until desired effects are achieved.

Step 5 Design Entourage Effect: Consider beta-caryophyllene, limonene, and beta-myrcene. Also, while peppermint oil does not occur in cannabis, it is a compilation of several individual terpenes, some of which also occur in cannabis, such as limonene, pinene, and cineol.

Step 6: Keep careful notes on your progress, use and adjust the pain scale on the worksheet in chapter 5 to your primary symptom and measure effects on it before and after usage. Revisit the scale at different times to understand the flow of effects in terms of duration.

Cough

Coughing is one of the top 10 reasons people seek out a physician's help and perhaps the most common affliction of people, and it remains one of the most vexing problems for patients and physicians alike. To date, there is no cure for coughs (or colds or flu) within the orthodox model of medicine.[1]

For the past two hundred years of modern medicine, opiates such as morphine or codeine have been used to treat a cough, with some degree of relief. However, especially in light of the recent opioid epidemic, the possibility of addiction or abuse, and common adverse effects ranging between constipation and respiratory arrest, fewer physicians feel inclined to use them.

Over-the-counter drugs containing dextromethorphan such as TheraFlu, Vicks, NyQuil, or Robitussin are commonly used with some degree of success, but a review of available studies conducted on the efficacy of dextromethorphan showed difficulty in producing clinically significant effects.[2] There are a number of other over-the-counter drugs such as Actifed or Sudafed that may suppress a cough reflex. However, they also contain speed-like substances (e.g., amphetamines) that can change your emotional state, make you anxious and jittery, keep you from sleeping, and raise your blood pressure and your heart rate.

Similar to a common cold or a flu, the major cause of an infection-based cough is a viral infection. Antibiotics target bacteria only and thus are ineffective in viral infections[3] and cause the destruction of beneficial intestinal flora. Single-ingredient anti-histamines may provide some minor relief in cases of allergy-based cough. Thus, in general, the need for relief from coughing remains largely unmet.

A word about coughing: A short productive cough (bringing up material) is the lungs' way to clean themselves and thus is a beneficial and necessary function. A chronic cough is an indication of an underlying condition (e.g., COPD, emphysema) or a maintaining cause (e.g., infection or inflammation due to allergens, toxins, or microbes; lowered immunity) and should be evaluated by a physician.

Warning: While most coughs will simply take their course and are by nature self-limiting, certain complications or conditions may warrant urgent treatment, especially if symptoms include shortness of breath, chest pain, cough-induced insomnia, fainting, vomiting, incontinence, hernias, nausea or vomiting, pale and sweaty skin, prolonged or very high fever, or tissue damage of the rib cage.

Cannabinoid-Based Research Highlights Relevant to Cough

Cannabis has proven anti-inflammatory, antispasmodic, and bronchodilating properties, all of which may play a part in producing the therapeutic impact of the herb as a potential antitussive (a drug used to relieve coughing).

As early as 1976, scientists experimenting on anesthetized cats discovered that THC administered intravenously to two test groups of cats (at dosages of 0.78 mg/kg and 1.84 mg/kg) had cough-suppressing abilities similar to that of codeine-PO. On the other hand, no such effects were noted in the test animals given CBD or CBN even at dosages of up to 10 mg/kg.[4]

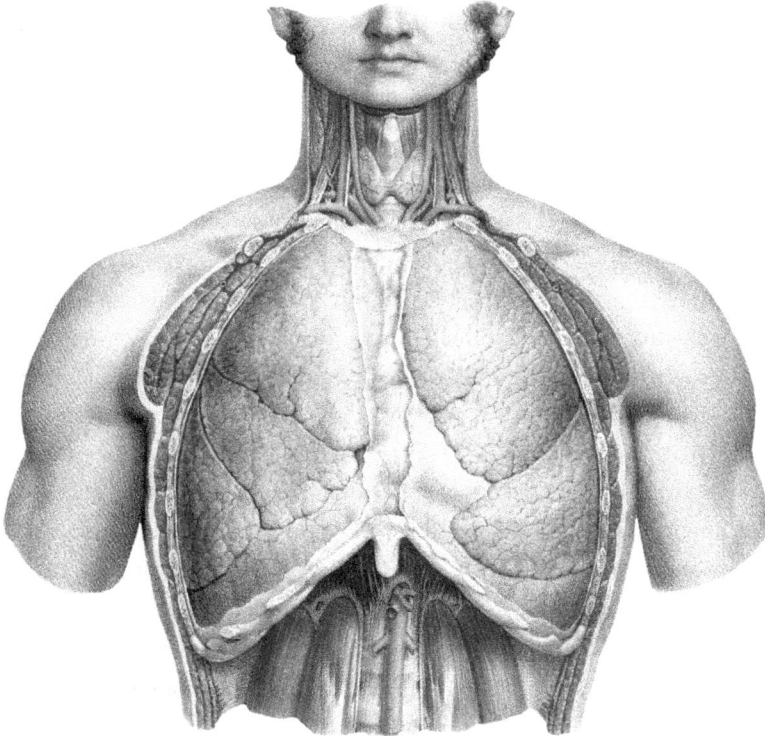

Appendix, Graphic 4

While the early findings of the antitussive properties of THC were confirmed in later experiments, the role of CBD as a potential remedy for cough via CB2 activation was revisited by British researchers in 2003. Indeed, it was shown that various synthetic cannabinoids reduce coughing by inhibiting vagus nerve activation. Researchers posited that developing a CB2-based anti-cough medication devoid of psychotropic (CB1 induced) effects may provide a novel and safe antitussive in cases of chronic coughs.[5]

An analysis conducted in 2006 by the University of California indicated that the immediate impact of smoking cannabis is bronchodilation. The abstract filed with the US Patent Office states: "The invention discloses the existence of cannabinoid receptors in the airways, which are functionally linked to inhibition of cough. Locally acting cannabinoid agents can be administered to the airways of a subject to ameliorate cough, without causing the psychoactive effects characteristic of systemically administered cannabinoids. In addition, locally or systemically administered cannabinoid in-activation inhibitors can also be used to ameliorate cough. The present invention also defines conditions under which cannabinoid agents can be administered to produce antitussive effects devoid of bronchial constriction."[6] Researchers hypothesized that their invention works via CB1 receptors.

 Additionally, a systemic review in 2007 of available studies conducted by researchers from numerous institutions confirmed that short-term exposure to cannabis produces bronchodilation.[7]

Researchers expressed concern that the long-term use of the plant might contribute to respiratory symptoms such as coughing, the review (by the University of California) acknowledged that these symptoms might also be caused by other factors, such as tobacco. This possibility is highlighted by the discoveries made in a 2009 Vancouver study: "Smoking both tobacco and marijuana synergistically increased the risk of respiratory symptoms and COPD (chronic obstructive pulmonary disease). Smoking only marijuana was not associated with an increased risk of respiratory symptoms or COPD."[8]

While it was well established that cannabis was able to produce bronchodilator, anti-inflammatory, and antitussive activity in the airways, which of the plant's many constituents were responsible was still subject to debate. An animal study conducted on guinea pigs in 2015 sought to clarify which major cannabinoids and by what specific pathways cannabis was able to elicited its antitussive effects. Delta(9)-tetrahydrocannabinol, cannabidiol, cannabigerol, cannabichromene, cannabidiolic acid, and tetrahydrocannabivarin were tested and compared. Results of the experiment showed that only Delta(9)-THC demonstrated effects on airway hyper-responsiveness, anti-inflammatory activity, and antitussive activity in the airways.[9]

In 2016 a group of British researchers conducted a laboratory experiment that broadened the understanding of how cannabinoids (in this case THC) are thought to therapeutically influence bronchoconstriction and airway inflammation common during acute or chronic phases of a number of airway diseases. Results showed that CB1 and CB2 receptor-like immunoactivity was found in bronchial epithelial cells, confirming the previous hypothesis of possible CB1-induced antitussive effects, but this is contrary to earlier study results indicating a possible novel therapeutic role for CB2-initiated action.[10]

An international group of scientists conducted an animal experiment in 2017 examining another aspect of antitussive pathways of the endocannabinoid system—that of enzyme-based degradation. Fatty acid amide hydrolase (FAAH) is an enzyme that breaks down endocannabinoids such as the body's own anandamide, which like THC binds with CB1 and CB2. It was hypothesized that enzyme FAAH inhibition would lead to an increase of the bioavailability of anandamide analgesic and anti-inflammatory effects relevant to upper airway inflammations and subsequent coughs. And, indeed, the guinea pigs responded to an elevated level of endocannabinoids with a reduction in cough as posited.[11]

Discuss with your licensed and cannabis-experienced physician if the following considerations for specific chemotypes, sub-ratios, forms, dosage, terpenes, or other plant constituents as well as their respective summaries are safe and of potential therapeutic value for you.

Chemotype-Based Considerations for Cough

The studies reviewed here discovered that specific cannabinoids (primarily THC and CBD) and the body's own anandamide with its enzymatic degradation pathways were involved in the amelioration of coughing albeit achieving results by different mechanisms. CB1 and CB2 activation via THC produced results. CBD reduced cough reflex by inhibiting vagus nerve activation. And finally, reducing the enzyme that breaks down anandamide (e.g., via CBD) produced an antitussive effect in the test animals. Thus all three chemotypes may provide relief (depending on the underlying cause) for coughing, especially if THC is substantially present.

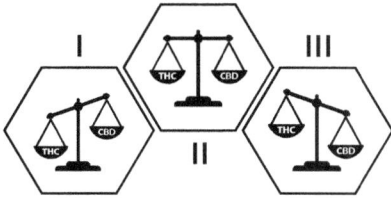

Sub-Ratio Considerations

Not enough data exist to make discerning decisions about sub-ratios. However, when using a cannabis chemotype I flower or product, you may want to consider starting low and incrementally increasing THC dosages, with the idea of balancing the goal of achieving the desired effect while avoiding any adverse reactions.

Form Considerations

Any inhalation of burned material is likely to induce coughing. Thus inhaling smoke may not be advisable. However, I have seen people with a cold-based cough inhale cannabis who got quick relief once the initial irritation had passed. If you have decided that you want to inhale cannabis, consider reducing additional irritation from smoking by using a vaporizing device that produces cannabinoid vapors while minimizing actual burned materials. Also, you may want to look for inhalers or sublingual spays for metered applications.

For longer-lasting antitussive results, an ingestible version (e.g., tinctures) might be of help. However, to date, no studies provide any results or guidance. As always, exercise caution when using carboxylated (psychoactive) ingestible forms of all three chemotypes unless they are devoid of or contain only a negligible amount of THC. The reader is reminded that ingestible cannabis-containing products take much longer to produce effects and that less is needed to cause similar effects compared to inhaled forms.

Some patients have reported success in producing anti-inflammatory and immunomodulating effects by using freshly cured flowers with most cannabinoids still in their acid form, thus engaging their raw form pathways to realize effective antitussive effects, especially in cases of longer-lasting cold- or flu-based coughs. Each would chop freshly dried cannabis flowers into finer material small enough to fill gelatin capsules and swallow them with water. Patients, especially those with compromised immune systems, ought to make sure to consume only tested plant matter devoid of toxic chemicals and mold. The latter method has not been tested or studied to date.

Dosage Considerations

These are based on those used in the human trials listed in chapter 3, section THC

THC micro dose: ~0.1 mg to 0.4 mg

THC low dose: ~0.5 mg to 5 mg

THC medium dose: ~6 mg to 20 mg

THC higher dose: ~21 mg to 50+ mg

Designing an Entourage Effect for Cough

Eucalyptol is used in a number of medicinal products aimed at mitigating symptoms of cough, nasal congestion,[12] and asthma,[13,14] and **pinene** has been found to dilate the bronchi (bronchodilation) in human subjects, even at low concentrations.[15] Some patients find it helpful to put a small amount of the terpene (or an essential oil rich in these terpenes) on their wrists to open the upper airways by scent exposure, thus creating a potential synergistic effect with other ingested or inhaled plant constituents.

Adverse Effects and Contraindications with Pharmaceutical Drugs

While the understanding of the interaction of cannabinoids and pharmaceutical drugs is in its infancy, it is understood that cannabinoids (phyto-, synthetic, and endo-) are molecules that are modulated (acted upon) by the superfamily of CYP450 enzymes. Thus the interaction between pharmaceutical drugs, endocannabinoids, phytocannabinoids, or synthetic cannabinoids at the CYP site produces the potential for an increased risk of treatment failure or drug toxicity as well as the possibility for creating drug synergies (e.g., using fewer chemotherapy agents to produce similar apoptotic effects).

Dextromethorphan (DXM) is an ingredient in a number of over-the-counter cough-suppressing drugs (e.g., Vicks, Robitussin, Benylin Dry Coughs), which is metabolized by the enzyme CYP2D. Research has shown that CBD is a potent inhibitor of CYP2D6[16] which increases the bioavailability of DXM and with it all of its potential effects (therapeutic and adverse alike).

Additionally, any cold, flu, or cough medicine containing ingredients that produce sedative or central nervous system depressive effects (e.g., opioids such as Robitussin with codeine) ought to be taken with caution. Concomitant use with cannabis may enhance their sedative effects, including hypotension (low blood pressure).

Summary: Cough

Step 1 Chemotype: Research suggests that all three chemotypes may provide relief (depending on underlying cause) for coughing, especially if THC is substantially present.

Step 2 THC:CBD Ratio: Discuss with your cannabis-experienced health care provider if starting with 1:4 ratio (2.5 mg THC:10 mg CBD) may be a good starting point for you.

Step 3 Choose Preferred Form of Inhalation: If your preferred form of medicine delivery is inhalation, consider using a vaporizer to minimize the irritating effects of smoke. Inhalation of cannabis flower provides quick but relatively short-lasting effects (about two hours). Consider an ingestible form when a longer duration of effects is needed.

Step 4 Optimal Amount: Consider, especially if you are concerned about cognitive changes, discussing with your physician if starting with a low amount of THC (e.g., 2.5 mg) is a good and safe option. Or for those with significant concerns about psychotropic effects, consider a micro-dosing approach (for more information go to chapter 5, step 4).

Step 5 Design Entourage Effect: Eucalyptol, pinene.

Step 6: Keep careful notes on your progress and use a modified pain scale on the worksheet to measure effects on your cough before and after inhaling. Revisit the pain scale at different times to understand the flow of analgesic effects in terms of duration.

Skin Conditions

One of the most common and prolific skin conditions, acne (like most chronic illnesses) has multiple underlying pathologies such as stress, poor diet, hormonal imbalances (testosterone tends to increase and estrogen decrease sebum production), inflammation, genetic predisposition, overproduction of sebum (oily secretion), and excessive proliferation of sebocytes (cells that produce sebaceous glands). Orthodox medicine has no cure, and effective treatment is wanting.

Sebum is produced by sebaceous glands, which are unevenly embedded throughout the dermis (middle) layer of the skin. As people who deal with acne know all too well, sebaceous glands tend to be highly concentrated in the face, scalp, and back (where acne tends to proliferate); there are none on the soles of your feet or in the palms of your hand. Hair follicles also pass through sebaceous glands, and thus overactive glands contribute to greasy hair and scalp.

When balanced, however, sebum functions to protect the skin and hair. Sebum waterproofs the skin and protects from the loss of moisture by providing a thin layer of oily substance. Additionally, sebum is slightly acidic and thus capable of neutralizing many environmental toxins (which tend to be alkaline in nature); it also has antimicrobial properties and thus represents a first line of defense against certain fungi, viruses, and bacteria.

Depending on where the cells are located in the layers of the skin, excess sebum can lead to whiteheads (located beneath the skin) and blackheads (on the surface of the skin) as well as pimples, zits, or pustules commonly referred to as acne.

Too little sebum and you have dry, itchy, or cracked skin, while too much may lead to the clogging of sebum and dead skin cells, which increases cellular pressure and may cause rupture—releasing sebum and other content into the surrounding tissue, causing acne. Patients with acne may also be at risk for secondary bacterial infections, which may prolong the healing process.

Cannabinoid-Based Research Highlights Relevant to Acne

Numerous endocannabinoid (originating from within) receptors (CB2, TRPV1, TRPV2, TRPV4) have been found in the skin and thus may present multiple new and simultaneous targets for the treatment of acne.[1,2] For instance, these naturally occurring cell receptors have been shown to balance cutaneous (skin) cell growth and differentiation (the process by which a less-specialized skin cell is transformed into one with a very specific function),[3,4] to balance sebum production,[5] and to induce systemic and localized anti-inflammatory effects,[6] all of which play a role in acne.

Additionally, the endogenous cannabinoids anandamide and 2-AG are both produced in human sebaceous glands and have demonstrated the ability to modulate their growth via CB2 signaling,[7] which in turn regulates gene expression involved in lipid synthesis via PPARSs,[8] resulting in reduced sebum production. Thus finding a way to amplify endogenous anandamide may be beneficial to shift the emotional environment away from stress and toward a more positive affect.

CBD inhibits the enzyme that breaks down anandamide, thus increasing the bioavailability of this endocannabinoid in the body, increasing the duration with which it gently activates CB1 and CB2 signaling that induces mood improvements (think stress reduction), and reducing sebum production via CB2/PPAR activation.

In addition, CBD has shown an ability to normalize and balance pro-acne molecules by reducing lipid production in sebocytes as well as by reducing sebocyte proliferation via pathways not modulated by CB1 or CB2.[9]

Research has demonstrated that tumor necrosis factor-alpha (TNF-alpha), a pro-inflammatory cytokine, is involved in acne pathogenesis.[10] CBD produces universal systemic anti-inflammatory actions. More specifically, researchers found that CBD prevented the increase of TNF-alpha.[11]

Furthermore, study results show that CBD has strong sebostatic actions mediated via TRPV4[12] and that modulation of CB2 produces an inhibition of sebum/lipid production.[13]

As we look at these research results collectively, a more complete and integrative approach to healing acne emerges.

Discuss with your licensed and cannabis-experienced physician if the following considerations for specific chemotypes, sub-ratios, forms, dosage, terpenes, or other plant constituents as well as their respective summaries are safe and of potential therapeutic value for you.

Chemotype-Based Considerations for Acne

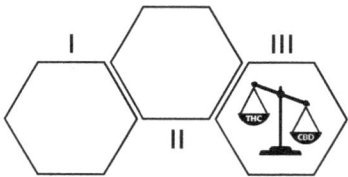

Data suggest that CBD has a positive localized effect in the skin where acne develops and thus seems uniquely positioned to address a number of the underlying causative pathologies of acne. More specifically, CBD's combined lipostatic, sebum anti-proliferative, and general and skin-specific anti-inflammatory effect has potential as a therapeutic agent for the treatment and prevention of acne. In addition, CBD has known therapeutic abilities in reducing certain moods commonly associated with acne such as anxiety or depression. Instead CBD confers mood improvements without euphoria or changes in cognitive function.

Thus research results suggest considering a chemotype III (high in CBD, low or negligible THC) for the prevention or treatment of acne and its commonly associated negative affect (emotions). Note that since THC is a partial agonist at CB2 and thus upregulates sebum production, current research would suggest not using a chemotype I with a relatively high THC concentration. If you need more THC for other reasons such as stress management (relaxation), consider using a chemotype II with a nearly balanced THC to CBD ratio and a relatively low concentration of THC.

THC:CBD Sub-Ratio (Topical): Consider discussing with your health care provider if a ratio ranging from 1:10 to 1:20 (THC:CBD) is a good starting point for you.

THC:CBD Sub-Ratio (Systemic): Consider exploring with your physician if a range of options between roughly 1:3 to 1:20 (THC:CBD) is safe for you.

Form Considerations

If you are opting for an ingestible form, consider choosing a product that does not contain processed or artificial sugar (e.g., soda, energy drinks, alcohol). Sugar is inflammatory. The more sugar you ingest, the more insulin and insulin-like growth factors your body makes. An increase in either can aggravate acne. Furthermore, avoid edibles that contain processed foods. Processed foods (e.g., fast foods, white bread, hormone-treated meat) contain compounds difficult to digest, which stress the body's pathways of elimination, including that of the skin. Avoid refined gluten grains (esp. wheat commonly treated with glyphosate, a compound in the weed killer Roundup), dairy products that are treated with growth hormones and can alter a healthy and natural hormone profile. Avoid trans fats (e.g., hydrogenated fat), which can also be highly inflammatory in nature, difficult to digest and eliminate. Instead try to increase your omega-3 fatty acids, which have been shown to reduce acne.[14]

If you are concerned about secondary infections, you might want to consider a topical product containing an alcohol-based cannabis chemotype III (CBD only, or much more CBD than THC). Many people find it helpful to use a clean Q-tip or sterile cotton swab to apply the topical to the affected area.

Dosage Considerations

These are based on the dosages used in the human trials listed in chapter 3, section CBD.

A low dose of CBD lies between 0.4 and 20 mg.

A medium dose of CBD lies between 20 and 100 mg.

A high dose of CBD lies between 100 and 800 mg.

Guiding Questions

Consider fortifying your cannabis-based product with limonene if your answer is yes to the following questions.

Is stress a trigger for me? Is my appearance with acne producing social anxiety? Is acne associated with loss of self-esteem? Does a breakout increase my negative self-talk? If stress is a known or suspected factor in the development of acne, consider mindfulness techniques such as meditation, tai chi, or yoga, which have demonstrably led to increased anandamide levels naturally.

Designing an Entourage Effect for Acne

Like THC (Ki of 35 nM), beta-caryophellene binds to CB2 (Ki of 155 ± 4 nM),[15] which means that beta-caryophellene has about a fifth of the strength of THC to initiate the same biological responses and effects such as balancing immune responses, triggering anti-inflammatory actions, and providing additional protection from oxidative stress while not being too strong to initiate CB2-associated increases in sebum production.

Chemotype III strains rich in (E)⊠BCP may additionally reduce sebum production via CB2-initiated PPAR activation of genes that code for specific proteins involved in aspects of lipid synthesis.

In addition, combining a chemotype III with the terpene limonene in the treatment of acne might allow the CBD-induced anti-acne effects be supported by its anti-inflammatory[16] and antioxidant properties.[17]

Also, if social stress produces anxiety or where a negative self-image results in depression, consider the following. Studies have shown that using limonene-fortified scents could produce anti-stress effects via modulation of parasympathetic functions,[18] could normalize neuroendocrine levels and immune function, and thus was rather more effective than antidepressants,[19] and experiments conducted on mice suggest that the scent reduced anxiety without any observable adverse effects.[20]

In addition, limonene-rich essential oils have been shown to inhibit the primary bacteria thought to produce Propionibacterium acnes.[21]

Further, alpha-terpineol has been shown to enhance the permeability of skin to lipid-soluble compounds,[22] which may help create a topical entourage effect with lipophilic cannabinoids. For instance, if a patient would like to employ the anti-acne effects of the lipophilic cannabinoid CBD, combining it with a small amount of terpineol may enhance localized bioavailability and better possible therapeutic results. As always, when working with essential oils, test a tiny amount on your forearm to make sure you are not allergic to the oil.

Additional Considerations

Echinacea spp. have been found to contain the compound N-alkylamide, which is a partial agonist at CB2 with about one third the strength of THC at the same receptor type[23,24] and may be considered to produce a similar effect as (E)⊠BCP.

Adverse Effects and Contraindications with Pharmaceutical Drugs

While the understanding of the interaction of cannabinoids and pharmaceutical drugs is in its infancy, it is understood that cannabinoids (phyto-, synthetic, and endo-) are molecules that are modulated (acted upon) by the superfamily of CYP450 enzymes. Thus interaction between pharmaceutical drugs, endocannabinoids, phytocannabinoids, or synthetic cannabinoids at the CYP site produces the potential for an increased risk of treatment failure or drug toxicity as well as the possibility of creating drug synergies.

Common pharmaceutical drugs used in the treatment of acne include antibiotics (tetracycline, erythromycin, doxycycline) and the application of hormone-altering compounds (e.g., contraceptives, spironolactone) or isotretinoin (for severe acne), for example.

Erythromycin is an inhibitor of CYP450 (1A2) and potentially increases the effects and duration of effects of the antibiotics in the human body.

Summary: Acne (topical application)

Step 1 Chemotype: Consider a chemotype III low in THC and rich in CBD.
Step 2 THC:CBD Sub-Ratio: Consider starting with a ratio ranging from 1:10 to 1:20 (THC:CBD).
Step 3 Choose Your Preferred Form: Consider an alcohol-based topical.

Step 4 Optimal Amount: When you are using a topical, you are delivering a strong effect to a very small area of your skin (the pimple).

Step 5 Design Entourage Effect: Consider finding a strain rich in beta-caryophellene and/or limonene.

Step 6: Keep notes of progress (or lack thereof) regarding your symptom reduction.

Summary: Acne (systemic application)

Alternatively, what you might do if you prefer to add a systemic approach for mood improvements.

Step 1 Chemotype: Consider a chemotype III (low THC:high CBD) to employ the anxiolytic and positive mood-elevating properties of CBD.

Step 2 THC:CBD Sub-Ratio: Your typical range of options lie roughly between 1:3 and 1:20 (THC:CBD). Let us look at one example using inhalation as a method of delivery and see how much actual THC and CBD from each end of the likely spectrum you would actually take. If you choose a 1:3 (THC:CBD) ratio with a 5% CBD content, then a sample of 50 mg of cannabis flower would contains 0.8 mg THC and 2.5 mgCBD (5% × 50 mg ÷ by 100% = 2.5 mg CBD; 2.5 mg CBD ÷ 3 = 0.83 mg THC). In contrast, if you choose a ratio of 1:20 and start with 5% CBD, you would get the same amount of CBD (2.5 mg) but your THC amount would reduce to a twentieth of that used in the previous example about 0.13 mg THC, significantly less, thus reducing all THC-induced effects.

Step 3 Choose Preferred Form: Consider inhalation, ingestion, or sublingual forms devoid of ingredients that might trigger acne.

Step 4 Optimal Amount

These are based on the dosages used in the human trials listed in chapter 3, section CBD.

A low dose of CBD lies between 0.4 and 20 mg.

A medium dose of CBD lies between 20 and 100 mg.

A high dose of CBD lies between 100 and 800 mg.

Step 5 Design Entourage Effect: Depending on your priority for either an increased synergy of anti-inflammatory action or mood elevation, find a strain rich in beta-caryophellene and/or limonene, respectively.

Step 6: Keep careful notes on your mood (consider using the modified pain scale to measure). If your mood improves and you are happy with your results, continue with your regimen. If your mood gets worse, reduce your THC intake by switching to a lower THC concentration or a lower ratio, or if not already using it, switch to a chemotype III. For additional anti-inflammatory fortification, consider adding *Echinacea spp.* to your regimen.

Forms of Cannabis

What Are the Various Forms of Cannabis You May Encounter and How Is That Relevant?

Cannabis-based products can be consumed in many different forms. Choosing the form most appropriate for you is an important step in taking control over the effects in terms of symptom reduction or mood improvements you would like to achieve. The key difference among these products centers on composition of cannabis constituents, THC concentration (which can vary greatly), and the safety of the end product, including production methods used. The most common forms can be divided into four groups: inhaling, ingesting, sub-lingual, and topical.

Using the example of John, an imaginary patient who needs 5 mg THC to reduce his symptoms, we will demonstrate how much of one of these products you would actually need to obtain the equivalent of 5 mg THC. Knowing these differences will allow you to make more informed decisions about meeting your specific goals and avoiding adverse effects. For additional information on calculating the concentration of THC, go to chapter 5.

What Do I Need to Know About Inhaling Cannabis-Containing Products?

If you wish to inhale cannabis, you have various options to consider, each of which delivers a significantly different concentration and therefore different amounts of plant constituents and the respective effects they engender. First, do you want to simply smoke a "joint" (a cigarette rolled with dried cannabis flowers and leaves), use a water pipe (bong), or use a vaporizer, vape pen (similar to an e-cigarette), pipe, or dabbing rig?

Second, you have a choice about what type of cannabis-containing product you would like to inhale—dried flower/leaves or concentrates such as kief and hash (solid hash or water hash aka bubble hash), or maybe further concentrated oils such as butane-extracted hash oils (BHO), or carbon dioxide (CO_2) extracted oil, which uses a number of monikers such as crumble, budder wax, sap, pull and snap, or a live resin (using fresh, frozen cannabis flowers).

With this many distinctions that underly the methods that produce the various looks, textures, and actual plant constituents, it is easy to get lost in industry jargon. And, while these differences and details can matter, the actual amount of THC is still of critical importance no matter what specific product you end up using, and thus is is a major factor for dosing.

Dried Flowers/Leaves

These are just what the name implies. Once cannabis is harvested, the plant's fan leaves are removed and the smaller leaves are trimmed. Think of it as giving the flowers a careful haircut to where only the denser, center mass of the flowers (the buds) remains. The flowers are now hung up or, in the case of loose buds, shelved and slowly, gently dried. When you visit a dispensary, you will see shelves of various buds with many exotic-sounding names with different textures, sizes, densities, and scents. Remember at this point your cannabis flower is not yet activated (i.e. decarboxylated). Furthermore, once activated the strength of cannabis flowers, buds, or leaves to produce psychotropic (and a number of specific medical) effects depends on the percentage-by-weight concentration of THC. Let us use a few examples examples of already decarboxylated cannabis flowers:

If 100 mg of activated flower contains 7% THC, that sample contains 7 mg THC.

For John to meet his target dose of 5 mg THC, he would need to consume about 71 mg of that flower sample. (5 mg THC x 100 mg flower ÷ 7 mg THC = ~71 mg flower)

If 100 mg of activated flower contains 14% THC, that sample contains 14 mg THC.

For John to meet his target dose of 5 mg THC, he would need to consume about 36 mg of that flower sample. (5 mg THC x 100 mg flower ÷ 14 mg THC = ~36 mg flower)

If 100 mg of activated flower contains 21% THC, that sample contains 21 mg THC.

For John to meet his target dose of 5 mg THC, he would need to consume about 24 mg of that flower sample. (5 mg THC x 100 mg flower ÷ 21 mg THC = ~24 mg flower)

Classical Cannabis Concentrates

Kief: This is one of the oldest forms of cannabis concentrate, dating back to ancient history. It is a sticky, fine powder (dried resin, sifted powder, crystal, pollen) with a golden to light brown (sometimes grayish) color that emerges naturally in the drying process, with an abundant and intense aroma. If you were to place a white cotton cloth beneath your drying cannabis flowers, within a few days you would notice the dried resin building up, as some of the trichomes, the hair-like appendages of cannabis that contain the plant's active constituents (e.g., cannabinoids, flavonoids, terpenes), dry, break, and fall down. Separation and collection of the resin is usually accomplished by using a sieve or ice water. Kief is smoked in the form of a cigarette or with the use of a vaporizer.

As a rule of thumb, kief may constitute approximately one quarter of the weight of a cannabis flower, but the strength of kief will be about double that of the cannabinoid content of the flower from which it was gathered and compacted.

If 100 mg of flower contains 7% THC, that sample contains 7 mg THC; 25mg of kief (100 ÷ 4 = 25) from that flower sample will contain about 14 mg THC. (7 mg THC × ~2 = 14 mg THC). For John to meet his target dose of 5 mg THC, he would need to consume about 9 mg of kief from that flower sample: (5 mg THC × 25 mg kief ÷ 14 mg THC = ~9 mg kief).

If 100 mg of flower contains 14% THC, that sample contains 14 mg THC; 25 mg of kief (100 ÷ 4 = 25) from that flower sample will contain about 28 mg of THC: (14 mg THC × ~2 = 28 mg THC). For John to meet his target dose of 5 mg THC, he would need to consume about 4.5 mg of kief from that flower sample. (5 mg THC x 25 mg kief ÷ 28 mg THC = ~4.5 mg kief).

If 100 mg of flower contains 21% THC, that sample contains 21 mg THC. 25 mg of kief (100 ÷ 4 = 25) from that flower sample will contain about 42 mg of THC. (21 mg THC x ~2 = 42 mg THC). For John to meet his target dose of 5 mg THC, he would need to consume about 3 mg of kief from that flower sample. (5 mg THC x 25 mg kief ÷ 42 mg THC = ~3 mg kief).

Hash (Hashish): This is a version of kief that depends on a method of extracting that presents as solid (when heated and pressed), or soft and paste-like (when concentrated using ice water). Hash is smoked rolled into a cigarette, with the use of a vaporizer, or by using a pipe. Concentrations of THC in hash should mirror those estimations calculated for kief.

Extraction-Based Concentrates

New forms of concentrates have become available and popular, bringing a shift in extraction methods and inhalation mechanisms. Instead of concentrating dried trichomes via a sieve (e.g., kief, hash), specialized extraction methods are employed to concentrate trichome content.

Many extraction-based concentrates center on increasing THC yield. As a result, some of today's concentrates approach 90%+ THC concentration. For some patients with certain types of pain, cancers, or chemotherapy-induced nausea and vomiting, such a high concentration of THC may be a necessity, and thus using an extraction-based concentrate may be a practical option. Because of its rapid-onset effects and ease of titration (for the experienced), using high concentrates has become a popular option for patients dealing with withdrawal symptoms associated with alcohol, opioid, or addictions to speed-containing drugs.

Chemical solvents such as hydrocarbons (butane, propane, hexane), carbon dioxide (CO_2), or alcohols (grain, isopropyl) are used in the extraction process. While these chemical agents efficiently dissolve and separate trichomes from the plant, they may also leave solvent residue or may accumulate and concentrate pesticides or toxins if cannabis source material was sprayed or otherwise exposed to toxic material. Researchers testing more than 57 concentrate samples discovered that over 80% of them were contaminated in some form.[1] Therefore it is vital to consume tested and verified organic products whenever possible in order to minimize negative impacts to your health.

Note: No long-term safety information on extracted concentrates such as those used in dabbing is yet available. However, a couple of studies may provide some context for discernment about the use of extracted concentrates.

A study from Switzerland examined the use of liquid refills for e-cigarettes enriched with cannabinoids including butane hash oil (BHO). The authors suggest that "cannavaping"—defined as the "vaping" of liquid refills for e-cigarettes enriched with cannabinoids, including BHO—appears to be a gentle, efficient, user-friendly, and safe alternative method for cannabis smoking for medical cannabis delivery. On the side of caution, researchers were concerned about potentially overheating the cannabis constituents, possible contamination, and potential for misuse.[2] In contrast, researchers at Portland University (2017) discovered that dabbing may be subject to significant concern due to the oxidative liability of terpenes when heated, which may expose the user to significant amounts of toxic by-products such as methacrolein and benzene.[3]

A study conducted by the State University of New York asked 357 cannabis users about their experiences with dabbing. Results suggested that while users themselves did not notice any more problems or accidents than with regular cannabis, they viewed it as significantly more dangerous. Further, "participants did report that 'dabs' led to higher tolerance and withdrawal (as defined by the participants), suggesting that the practice might be more likely to lead to symptoms of addiction or dependence."[4]

To make highly concentrated cannabis forms bioavailable via inhalation, temperatures in excess of burning a joint or using a normal vaporizer became necessary. So, before we look at extraction-based concentrates individually, it may help to first examine the ways in which they are heated and inhaled.

Enter "dabbing." The name dab or dabbing is derived from the way many of the extracted concentrates are inhaled. A specially constructed glass bong made in part from titanium or heat-resistant quartz able to withstand extreme temperature significantly in excess of 315°C (600°F) is preheated with a specialized heating device such as a "torch" or an "e-nail," the latter of which tends to be more user friendly and safer, and prevents overheating.

A tiny amount of concentrate is picked up with a utensil called the "dabber," which is used to navigate the tiny spaces on the "rig." Some people prefer to flavor or fortify their concentrate first, either for taste or for specific therapeutic application. To fortify your concentrate, a dab may be dipped into the terpene or terpene profile of your choice just prior to placing or "dabbing" the concentrate onto a tiny input container, or small bowl (think replacement of traditional herb bowl).

Dabbing with the entourage effect in mind: Good quality, organic, food-grade, and tested terpenes and popular strain profiles are available at certain dispensaries or for purchase online.

The heat quickly turns the concentrate into a vapor that is typically inhaled through a glass tube (think straw), creating a near-instant and often intense experience. Beginner be warned. The risk of overshooting your subjective therapeutic window is significant and real. Consider using the equivalence approximations (John's example) to help navigate the amount you are dabbing.

Other concerns: Using a dab rig involves the obvious risk associated with the proximity of very high-temperature flames and heated objects close to your face. The possibility of inhaling metal fumes, toxic dust, and residue from the combustion process is a risk you may be able to minimize by carefully cleaning your rig bowl. Because you are dealing with extreme heat, and depending on the dabbing rig, potentially unstable temperature changes may destroy your cannabis constituents or diminish the therapeutic potential.

The following are some of the specific extraction-based concentrates used for inhalation.

Butane hash oil (BHO): Butane is a colorless and flammable gas commonly found in cigarette lighters and as a propellant in spray cans. It is a petroleum product extensively used in the plastic and solvent industries. Hydrocarbons are not commonly used in the food industry, but recent trends in molecular gastronomy (creative cuisine) sometimes utilize edible or food-grade aerosol sprays containing butane or propane, in the processing of cakes or chocolate.

BHO is made by exposing cannabis or hash to butane. When done professionally, the gas and cannabis are mixed in a closed-loop extraction system whereby trichomes dissolve and separate. The concentrate is collected, the gas recycled and reused. The process can be safe and clean—that is, without explosions or contamination of the air or natural environment. However, when butane is used outside a closed loop, that is, released into the air, the risk of fire and environmental damage is significant and real.

BHO comes in various forms, often named by using the visual description that is most applicable. You can choose honey oil (viscous golden oil), shatter (glass-like substance with a golden color), or wax (a sticky, caramel-like appearance), for example. Key health concerns include residual presence of butane. Different states or municipalities issue various criteria measured in parts per million (ppm) regarding what residual quantity is considered safe. While some use an upper limit of 5,000 ppm, the City of Berkeley, California, for instance, has issued a comparatively low permissible amount of residual flammable solvents: They must not exceed 400 ppm (total of all solvents).[5]

If 500 mg of BHO contains 70% THC, that sample contains 350 mg THC. (500 mg BHO × 70% THC ÷ 100% = 350 mg THC). For John to meet his target dose of 5 mg THC, he would need to dab ~7 mg of BHO. (5 mg THC × 500 mg BHO ÷ 350 mg THC = ~7 mg BHO)

To date, no studies could be found examining the presence of terpene abundance in BHO concentrates.

CO_2-extracted concentrated oil: CO_2 use is a common application and additive used in food processing and preservation. We consume CO_2 in carbonated drinks. When we drink decaffeinated coffee, the lipophilic caffeine molecules have been dissolved and removed using supercritical CO_2-extraction methods. It is commonly used as a propellant in whipped cream.

Trees "inhale" CO_2 and exhale oxygen (O_2). We breathe in oxygen (O_2) and exhale CO_2. Both gases are part of the circle of life called air. However, CO_2 can have a negative impact as well. Higher concentrations of CO_2 cause dizziness and headache, and can impair awareness. Severe exposures, especially in small, poorly ventilated spaces, can lead to death by asphyxiation.

CO_2-extracted concentrated oil is made using supercritical carbon dioxide. In this process CO_2 is exposed to high temperature and pressure, which produces changes in state from a gas to a liquid, allowing for targeted extraction of specific plant constituents. As with the use of hydrocarbon in professional facilities, CO_2 is processed in closed-loop circuits to prevent the gas from causing any ill effects to people or the environment.

CO_2 oils usually appear more clear (free of chlorophyll) and vary in color between an amber and honey shade of gold-brown. They are also likely to be less syrupy than BHO and thus lend themselves more naturally to use in vape pens (similar to an e-cigarette). CO_2 oils are safe as long as the source material and machinery were pure and free of contaminants.

If 500 mg of CO_2 oil contains 80% THC, that sample contains 400 mg THC. (500 mg CO_2 oil x 80% THC ÷100% = 400 mg THC). For John to meet his target dose of 5 mg THC, he would need to dab about 6.3 mg of CO_2 oil. (5 mg THC × 500 mg CO_2 oil ÷ 400 mg THC = 6.3 mg CO_2 oil).

To date, no studies could be found examining the presence of terpene abundance in CO_2 concentrates.

Live-resin extract: This is produced by flash-freezing freshly harvested cannabis (eliminating the drying process altogether). It is a new method of extracting concentrates using cryogenic technologies. Live-resin extracts significantly increase terpene abundance and with it scent and taste, making designing your entourage effect an even more potent element in fine-tuning your therapeutic cannabis experience. It is still rare to find this product, and it is more expensive than other concentrates.

If 500 mg of live-resin extract contains 90% THC, that sample contains 450 mg THC. (500 mg live-resin extract × 90% THC ÷100% = 450 mg THC). For John to meet his target dose of 5 mg THC, he would need to dab about 5.5 mg of live-resin extract. (5 mg THC × 500mg live-resin extract ÷ 450 mg THC = 5.5 mg live-resin extract)

What Do I Need to Know About Ingesting Cannabis-Containing Concentrates?

You may know your history or remember from chapter 1 that for much of the 19th and 20th centuries the primary physician-prescribed delivery mechanism for cannabis was an alcohol-based tincture. It was effectively used to treat a variety of conditions. As , it is a time-proven tradition that has made a powerful comeback. Ingesting cannabis-containing products is once again an attractive and effective way for many patients to alleviate symptoms.

However, while ingesting cannabis can produce similar effects on the body, mind, and emotion, when inhaling, the time of onset and intensity of effects can vary significantly. Minor misunderstandings about how it works are the leading cause of adverse effects such as sensations of anxiety or panic, or an actual increase in the symptoms you are trying to alleviate. For instance, a common mistake is to assume that smoking 50 mg of cannabis flower produces the same effect as ingesting 50 mg of THC in the form of a tincture or an edible. By now the reader knows this not to be true, but it may be worth revisiting.

Understand that the form, in this case inhalation or ingestion, determines the amount of cannabinoids that you absorb, and the kind of cannabinoid metabolites (breakdown products) that become biologically active and enter your bloodstream. Thus, different amounts and different cannabinoid metabolites contribute to produce variations in biological and psychological effects. Knowledge about these variation allows you to produce the therapeutic effect you desire or prevent potential adverse effects you want to be sure to avoid. Here are the basic differences.

When inhaling, effects come on quickly. The benefit of feeling the effects within minutes allows you to stop when the desired effect is achieved (a process called titration) or to stop instantly if any adverse effects occur. Depending on each person and cannabis variables, effects induced by inhalation may last up to about two hours. For some patients trying to get a good night sleep, for example, a therapeutic window of two hours is not long enough. Here is where carefully used cannabis-containing tinctures or the products that contain them can make a real difference. Why?

When cannabis-based tinctures (or the products that contain them) are ingested, they are processed in a different way than when you inhaled them, which results in a much longer onset time for effects to take hold. However, once cannabis constituents are broken down in the liver-centered digestion process, they enter the bloodstream and make it past the blood-brain barrier (BBB). Depending on stomach contents and other variables, effects occur approximately one to two hours after actual ingestion. And, here is the difference to remember.

When THC passes through the liver, a portion of it is oxidized into an additional, but also psychotropic, metabolite called 11-hydroxy-THC, which has been shown to penetrate the BBB at a rate much faster than an ingested THC molecule by itself.[6] So, while it takes longer (up to 1–2 hours) for ingested THC to enter and peak in the blood and brain, once converted, both THC and 11-hydroxy-THC produce a synergy and thus rapid absorption, leading to sudden and often more intense changes in body and mind. So, once you know your optimal ingestible dose, you can enjoy the therapeutic effects for many hours on end. How long exactly? Much of it depends on variables such as your endocannabinoid tone or how much you took. Effects may last between 4 and 10 hours and longer if you took a dose causing significant unpleasant effects. Also, as concentrations of cannabinoids change over the course of your cannabis experience, effects may also shift accordingly. For instance, you may notice the onset of relaxing effects at hour 1.5 and lasting for 4 hours, you may notice a "high effect" for 6 hours, and a reduction of nausea and vomiting for a full 24 hours. This is why journaling your experience is one of the most empowering things you can do to turn your cannabis prescription into a precision medicine that serves you as best as possible. If you had some prior therapeutic experiences with inhaling cannabis, you may wonder if you can apply what you know about it to using an ingestible form? You may. Research suggests a ratio of ingestion to inhalation of about 1:6.[7] That is, for every 1 mg of ingested THC, about 6 mg of inhaled THC may be necessary to achieve a similar effect. The reader is asked to keep in mind that this ratio is an estimate, to be considered with appropriate caution. It was derived mathematically from averages of different study results, each with ranges that had large margins. So, especially when it comes to ingestion, cannabis-containing products start low and slow.

What Do I Need to Know About Sublingual Cannabinoid Delivery Systems?

Drugs can be administered by placing them under the tongue. Products that employ sublingual deliveries include oromucosal sprays (aerosols), strips, chewing gums, and nano (cannabinoid) emulsions.

The mucous membrane and vascular tissue beneath the tongue is made up of a dense population of capillaries (tiny blood vessels) that will quickly absorb any substance as long as it dissolves in saliva.

Pharmaceutical drugs are produced and packaged in various ways to make use of these natural pathways into the body. Examples include sublingual nitroglycerin sprays to quickly reduce angina-induced cardiac pain; the use "lollipops" of fentanyl, a potent opioid for the treatment of pain; or the use of Sativex, a full-spectrum cannabis-based drug, for the treatment of neuropathies associated with MS.

> What is metered *vaping*?
>
> Metered vaporization devices are beginning to appear on the market. Numerous states are planning to regulate metered dosing, matching pharmaceutical standards accurate to a tenth of a milligram. Keep in mind, however, that since responses vary greatly among individuals, even with metered dosages it would still be imperative to keep careful records of your progress in order to turn your cannabis prescription into precision medicine.

A drug delivery system that uses a dissolvable strip or one that sprays a metered (specific) dose of cannabinoids under the tongue (sublingual) has the obvious benefit of avoiding inhalation of carbon particles. But, perhaps most important, this method bypasses the digestive tract and liver, where THC is broken down and oxidized into various active and inactive components that make producing and sustaining specific effects more challenging.

As is the case with the pharmaceutical examples used above, sublingual delivery of cannabinoids enter the bloodstream by penetrating the tissue beneath the tongue. Penetration speed and efficiency is determined by how easily a molecule can navigate the watery oral environment and penetrate the tissue under the tongue. Cannabinoids are notoriously hydrophobic (water hating), and while some will make it through, they do not easily cross.

Numerous companies have been emerging and the benefits of specific formulations and delivery methods described as nano-forms (e.g., emulsions, spray, strip, gel) or nano-delivery technologies. Nano-products tend to use surfactants-based processing techniques, previously developed for the cosmetics industry, to make cannabis constituents more soluble in water and thus more readily absorbable. Some surfactants are nontoxic; others are not. In addition to using surfactants to make cannabis constituents act less hydrophobic (note "act," not "make"), ultrasound technologies are used to break down extracted cannabinoid droplets into smaller versions in the nanometer scale (1 billion of a meter or $1\times10{-9}$ m). It is posited that the ultrasound treated smaller droplets with the additional help of surfactants make cannabis constituents more bioavailable and thus make absorption more effective at lower dosages and at a lower cost.

What Are Some of the Other Forms I Am Likely to Encounter?

Where you are shopping for your cannabis-containing products determines access to variety. Some online purveyors and some patient-oriented dispensaries may also offer specific options such as suppositories (vaginal or rectal), transdermal patches, nasal sprays, or erotic lubes.

Terpenes and Your Health: Deeper Background and Context

What Is an Entourage Effect?

The term *entourage effect* was first coined by the Israeli researcher Shimon Ben-Shabat in a paper published in 1998.[1] In it his team describes a relationship between various endocannabinoids. One compound showed no activity alone but became biologically active in the presence of two other related compounds. Thus, over time, an entourage effect can be defined as one that enhances or magnifies therapeutic effects in body, mind, or emotion, produces balance between organ systems (homeostasis), enhances efficacy, or reduces adverse effects. For instance, the terpene myrcene enhances the sedative effects of THC, while CBD tames the psychotropic effect of THC.

In essence, the entourage effect is the biological basis for why full-spectrum cannabis is so significant. It is why stand-alone cannabinoid-based products, while having their place in medicine, will not be able to serve a great variety of different patient populations that need the complexity and multi-factorial effects only a whole plant medicine can provide.

Some people choose their cannabis flower with an eye (or nose) on its terpene profile. A visit to a dispensary that offers test results will be very helpful for making informed decisions. Others, for example, those fond of working with extracts and dabbing, have taken to dipping their dab into the terpene or terpene profile of their choice. Yet others simply put a drop of the terpene oil of their choice, or an essential oil that contains them, onto their wrist or neck for the duration of the inhalation process. Still others enhance their buds by misting a tiny amount of their favorite terpene onto the flower.

To better understand the often subtle but deep reaching effects of the entourage effect, it is helpful to look at those individual cannabinoids and terpenes that have a proven ability to enhance, balance, and protect in concert. It is also important to realize that many of these effects begin the moment our nose takes a whiff.

Terpene	Abundant in Essential Oils of
Caryophyllene	*Piper guineense* (White or Black Ashanti peppers), *Aframomum melegueta* (Grains of Paradise), *Cinnamomum tamala* (Indian Bay leaf), *Artemisia annua* (Sweet wormwood), *Copaifera* (Copaiba)
Limonene	*Citrus paradisis* (Grapefruit), *Citrus limon* (Lemon), *Citrus sinensis* (Orange), *Mentha spicata* (Spearmint)
Linalool	*Coriander sativum* (Coriander), *Ocimum basilicum* (Basil), *Lavandula spica* (Lavender)
Eucalyptol	*Eucalyptus species, Melaleuca alternifolia* (Tea tree oil), *Mentha spicata* (Spearmint), *Melaleuca cajuputi* (Cajuput)
Phytol	~30% of all chlorophyll-containing plant parts is phytol
Nerolidol	*Melaleuca quinquenervia* (Niaouli)
Myrcene	*Thymus serpyllum* (Wild Thyme), *Humulus lupulus* (Hops), *Elettaria cardamomum* (Cardamom)
Pinene	*Coniferae* (Conifers), *Pinaceae* (Pines), *Heterotheca shevockii* (Camphorweed)
Camphene	*Cinnamomum camphora* (Camphor tree/laurel), *Rosmarinus officinalis* (Rosemary)
Terpineol	*Satureja montana* (Mountain savory)
Terpinolene	*Microtoena patchoulii* (Chinese patchouli)
Borneol	*Heterotheca shevockii* (Camphorweed), *Kaempferia* (various species of ginger, e.g., *K. galanga*)
Geraniol	*Monarda*, various mint species, e.g., *M. fistulosa* (bee balm), *M. didyma* (bergamot)
Humulene	*Humulus lupulus* (hops)

Appendix, Graphic 5: Terpenes Commonly Found in Cannabis and the Essential Oils That Tend to Carry Them in Higher Concentrations

Exploring Scent as a Cornerstone of Health and Well-Being

How many scents can we distinguish?

Humans are able to distinguish among more than 1 trillion olfactory stimuli.[2] Think about that for a moment. We can sense the difference between more scents than there are stars visible in the night sky. In contrast, our sense of taste can distinguish only between sweet, sour, salty, bitter, and savory (umami) tastes. In fact, when we experience other tastes, we actually smell them as the scent fills our throat and reaches the olfactory bulbs in our sinus cavities.

Do the emotions of others have a scent?

We are accustomed to the idea that dogs have an amazing nose; however, we humans can be pretty effective when it comes to smell, too. Consider that we can even smell the way another person feels.

Researchers from Utrecht University, the Netherlands, discovered that humans can smell emotions in others. A double-blind study revealed that chemo-signals of fear and disgust provoked the same emotions in people exposed to them, demonstrating that these emotional chemo-signals operate outside conscious awareness.[3] It could be posited that this discovery is a biological basis in the formation of empathy, potentially explaining the phenomenon of "contact high."

Are there other ways we can pick up scent besides the nose?

In another fascinating discovery, scientists demonstrated that the human skin is able to smell. Prolific skin cells called keratinocytes actually contain at least five distinctly different olfactory receptors that when stimulated by the scent of sandalore (a synthetic sandalwood oil) initiate skin cell division (increased by 50%) and migrations (increased by 50%), both hallmarks of skin healing.[4]

In fact, scent may even play a critical role in procreation at its most basic level. Scent receptors have been found on the surface of human sperm cells. Sperm may be able to smell the shortest distance to the egg and determine their course by following their "nose."[5]

How does scent enter the body? How does scent affect the way we feel and act?

Scents can enter the body by inhalation, ingestion, and absorption through the skin. Once inside the human body, scent begins to produce measurable changes to body, mind, and emotion. Perhaps more than any other sense, the experience of scent can produce a variety of near-instant, and often intense, emotional experiences, depending on what memories or associations have been married to what scent. It can trigger vivid memories and feelings across the emotional spectrum, ranging from fear, hate, and disgust to trust, relaxation, and arousal. And this often occurs so rapidly that, unless we are very conscious of the process, it effectively robs us of a chance to discern before we act. We catch a whiff of something acrid or rotten, and we recoil without the benefit of reason, or using logic to determine a best possible response first.

Can scent produce biological, psychological, and behavioral changes?

As you might imagine, this direct connection between nose, brain, emotional experience, and behavior can be either detrimental or a healing balm to certain patients, such as those suffering from PTSD. This is a unique opportunity for patients and practitioners alike to match scents with specific symptom reduction and mood improvements. Medical history is rich with examples of using scent to soothe what ails us, and the traditions date back thousands of years. No lasting medical tradition is devoid of it. In the cannabinoid health sciences, this process can be, at least partially, explained by the "entourage effect." However, scent-induced synergistic effects are only one expression of the entourage effect. Other times cannabis constituents can shift homeostasis or balance between various human organ systems by canceling a specific effect or by strengthening another.

For instance, CBD may tame the psychoactive effects of THC, while the terpene myrcene makes THC more bioavailable by increasing the affinity and duration of CB1 signaling, thereby increasing its properties such as the feelings of heavy, deep relaxation—often referred to colloquially as the "couch lock" phenomenon.

Understanding the complex relations among scent, emotion, and behavior is an intricate and subjective affair. Have you ever noticed how difficult it is to describe a scent to someone who has never smelled it before? In addition, few if any people experience scent in exactly the same way. This is true for intensity, emerging memory, the resulting emotional experience, and the instant reaction, choices, or behavior associated with it. While one person might feel bliss when smelling freshly cut grass, another might feel an intense anxiety.

This subjective difference between people's experiences with a particular scent highlights the importance of involving your own "sensing" responses (physical, mental, and emotional) when working with cannabis-containing products. This is true in general and more specifically when designing the best possible entourage effect for you.

Also, consider this: When we experience emotions, scent-induced or by other means, the body instantly initiates the release of its own molecular counterparts. Fear increases adrenaline, pro-inflammatory cytokines, and cortisol. Bliss produces oxytocin, serotonin, and anandamide. You get the idea. But what you may not know is that many of these hormones (e.g., oxytocin, cortisol, adrenaline), neurotransmitters (e.g., serotonin, anandamide), signaling proteins (e.g., cytokines), and their corresponding emotions are modulated, in part or significantly, by the body's own endocannabinoid system. These modulations function in general to mitigate stress, boost energy levels, and improve moods. More specifically, research has shown that engaging the axis of scent and emotion, modulated via the ECS, can produce real and specific physiological changes that can be harnessed thus are relevant to various patient populations, especially those dealing with chronic conditions. For instance, it has been demonstrated that scent is able to reduce blood pressure[6-9] and anxiety,[10] as well as enhance brain power[11]—effects of high relevance to those suffering from hypertension, anxiety disorders, and Alzheimer's disease, for example.

At this point you might ask, how specifically does smelling something produce change in body, mind, and emotion? Scent compounds are lipophilic (fat-loving) and hydrophobic (water-hating) by nature. This simply means they readily interact with lipids (fats) but not or only very little with water. Membranes of human cells are composed of lipid layers. Thus scent compounds readily interact with a variety of cell types and by different means.[12] For instance, scent (in the form of plant-based terpenoids) can modulate neurotransmitters (e.g., serotonin, dopamine),[13,14] change enzyme bioavailability,[15] modify ionotropic acetylcholine receptors,[16] or bind with certain G-protein-coupled receptors.[17]

Many times people wonder how synthetic essential oils or terpenes differ from naturally produced sources. For the most part, science, looking at the precise, small, focused impact of a compound, does not distinguish between a molecule made by nature or one produced artificially. However, some people concerned with the spirit and soul of a plant may take a different view. Ultimately you decide which product you prefer.

The Scent of Cannabis—How Does a Subtle Whiff Make Such a Difference?

In all three cannabis species, terpenes and cannabinoids share creation (biosynthesis) pathways and locations of accumulation. Both are located in the small mushrooming, hair-like features or tiny outgrowths (trichomes) of the outer layer (epidermis) of the cannabis flowers and to a lesser degree on the leaves.

Trichomes can be seen with the naked eye and upon closer inspection with a microscope reveal a crystalline nature beneath the veneer of a sticky, shiny, glass-like appearance. Among their various functions in the larger biology of life, these compounds give each plant its unique odor, which both repels harmful pests such as mites and attracts beneficial insects such as ladybugs that devour aphids. In addition, trichome blankets function to protect against high temperatures, UV light, loss of moisture, and certain herbivores.

Terpenes are considered one of Mother Nature's most common building blocks, with more than 30,000 different varieties discovered to date.[18] Terpenes are also commonly referred to as terpenoids, which may produce some confusion. Allow me to clarify.

Early German organic chemists isolated the first terpenes from turpentine, a common solvent. These chemists considered the compounds to be the smallest building blocks (units composed of 10-carbon atoms) of this class of chemicals, and thus called them monoterpenes. The name survived, forcing future generations of chemists to be creative in naming smaller compounds as they were discovered.

The method of naming that resulted found a way to respect the historical origin. It cleverly uses the fact that terpenes are actually made of 5-carbon-unit building blocks (called isoprene units) and applies the Latin prefix of fractions to specify the various types or classes of terpenes. Thus, **monoterpenes** consisting of 2 isoprene units are counted as 1 (mono), **sesquiterpenes** consisting of 3 isoprene units are counted as 1.5 (sesqui), and **diterpenes** consisting of 4 isoprene units are counted as 2 (di), and so on.

In some papers and books, the distinction between terpenes and terpenoids is described as follows: Terpenes are simple hydrocarbons (composed of building blocks of 5-carbon units), while terpenoids are similarly composed of hydrocarbons (i.e., they are also composed of building blocks of 5-carbon units), and they contain an additional functional group, that is, a group of atoms that bestows specific properties to a molecule. For instance, the carboxyl group keeps THC in its acid form (THCA) and thus is characteristically non-psychoactive. However, while both designations are current and address certain aspects relevant to the intricacies of organic chemistry, for the purpose of this book the terms can be used interchangeably.

Synonyms commonly applied to terpenoids are the terms *volatile oils* and/or *essential oils* (EO). I have come to think of terpenoids as the essence of a plant's aroma. Essential oil of cannabis is produced most frequently by steam distillation. It is very rare and one of the most expensive essential oils. Cannabis EO is used in very diluted forms and often mixed with carrier oils for topical applications. It takes 1,000 kg of fresh flowers to produce 1.3 L of cannabis essential oil, or 1 kg to produce 1.3gm.[19]

By 2007, over 140 terpenes had been identified in cannabis species alone, and this number is constantly growing as new ones are discovered.[20] To date, terpenes discovered in cannabis fit into the three groupings (types) mentioned earlier (monoterpenes, sesquiterpenes, and diterpenes). And while most of them can be detected by the human nose, some do not concentrate in large enough amounts to be considered pharmacologically relevant. However, that does not mean they are irrelevant, especially in producing full-spectrum synergies that are perhaps too subtle to be consciously experienced. They too may still play a role in contributing to our journey toward optimal health.

So, how much or how little does it take to produce a noticeable or measurable effect?
Researchers posit that, based on current analytical limits and technological measuring capabilities, a threshold of terpene concentrations of 0.05% or higher is a reasonable mark to expect pharmacological effects in humans.[21] The terpenes produced by cannabis that have been found to exist in concentrations greater than 0.05% tend to be the following (see appendix, graphic 5).

> **Note to reader:** You may notice in the following pages that the molecular formula for specific terpenes (e.g., limonene and camphene) is the same ($C_{10}H_{16}$). This simply means that the molecules are isomers, that is, built from the same number of atoms but connected in a different spatial arrangement.

As we already know, these substances and scents have positive medicinal value in the existing tool kit, so we can intuit that the volatile oils, the terpenes, of cannabis share a number of effects relevant to human medicine and well-being. Selected study results examining the effect of individual terpenes on a variety of medical conditions are presented in the following pages. You may want to consider this summary, in addition to the material covered in chapter 5, step 5, as a guide to help create entourage effects and fine-tune your own unique course toward optimal health and well-being. As you will notice, we list terpenes (from top to bottom) based on the amount of scientific literature currently available for each. For some terpenes there is very little scientific literature (e.g., Humulene), while in other cases there is significantly more (e.g., beta-caryophyllene). Keep in mind that research interests tend to go in the direction of where there is significant excitement about new and effective treatment possibilities. However, this does not necessarily mean that a terpene with less study results is ineffective. It may just be a reflection of the focus of potential impact. In some cases, the focus may be narrow, while in other cases it is applicable to a wide variety of conditions or underlying pathologies relevant to numerous diseases. Or it may be that the full potential of a terpene is still waiting to be discovered.

As you review the following pages, you might want to note on the margins or a notepad or use the space provided in step 6, which terpene's actions match your need for symptom reduction and/ or mood improvements. Use your own notes/list as a guide to find or fortify your optimal synergy.

Once you discover a specific terpene profile that works best for you, perhaps you can grow the same strain under the same conditions and in the same environment, or look for a tested product that mirrors your preference most closely.

Also, flowers and most cannabis-containing products can be fortified by adding specific synergistic (certified organic) terpenes using a mister or dropper and either spraying it directly onto the flower or adding a precise amount to a tincture, oil, edible, or even putting a drop of an essential oil rich in it onto your wrist or neck, for inhalation of its aroma to enhance your overall cannabis experience.

Keep in mind that while monoterpenes dominate the olfactory experience and may represent up to 10% of fresh trichome content, by the time a flower has gone through the drying process and is in storage, the actual monoterpene yield of most dried flowers is significantly less than 1%, while sesquiterpenes such as beta-caryophyllene remain stronger in comparison.[22]

Specific Terpenes and Your Health: Deeper Background and Context

Monoterpenes: Borneol, Camphene, Eucalyptol (aka 1,8 Cineole), Geraniol, Limonene, Linalool, Myrcene, Ocimene, Pinene, Terpineol, Terpinolene

Borneol

Borneol is a minor terpene in cannabis and more abundantly found in turmeric, thyme, and ginger, for example. It is a compound frequently used in Traditional Chinese Medicine as an analgesic and was sold in pharmacies of old as a dry powder.

Summary of selected effects on the body: Antibacterial (middle ear infection), analgesia via TRPM8.

Researchers at the First People's Hospital of Jinin, China, conducted a human trial on 278 human patients suffering from a pus-producing **middle-ear infection (purulent otitis media)**. They divided the patients and treated one group with borneol-walnut oil and the other with a pharmaceutical counterpart, a neomycin compound (antibiotic). Over 98% of patients treated with the topical borneol-walnut oil (optimal concentration was 20%) were cured compared to 84% of those treated with the **antibiotic** alternative. Thus the terpene exhibited superior antibacterial action with no adverse effects.[1]

In another trial from China, a randomized, double-blind, placebo-controlled clinical study involving 122 patients with postoperative pain, demonstrated that topical application of borneol produced a significantly greater analgesic effect than the placebo. To learn how borneol induced **analgesia,** these researchers tested the compound on mice and found that the pain-relieving properties were almost exclusively mediated via TRPM8 receptors.[2]

Summary of selected effects on mind and emotion: None known.

Safety: Concentrated borneol ($C10H18O$) is a pale, white, lump-like solid with an intense camphor-like aroma.

Camphene

Camphene is a common scent and flavor additive in certain foods and fragrances. It is described as having a cooling, woodsy, fir-needle type of smell. I think of it as having a bland camphor-like scent. Camphene has a few unique medicinal properties.

Summary of selected effects on the body: Analgesia (via TRPV3), may reduce hyperlipidemia and related conditions (e.g., cardiovascular illness, cholesterol-based gallbladder stones).

Topically it can produce the sensation of coolness or warmth via ion channels (TRPV3)[1] and thus is used in **topical analgesia** balms.[2]

An experiment conducted on rodents demonstrated that camphene may play a role in reducing **hyperlipidemia**, a possible co-factor in developing **cardiovascular illnesses**.[3]

Camphene (along with cineole and pinene) is an ingredient in Rowachol (Rowatinex) capsules, composed of an essential oil-based drug available in Europe and designed to break up **cholesterol-based gallbladder stones.**[4]

Summary of selected effects on mind and emotion: None known.

Safety: Isolated camphene (C10H16) is a crystalline-like solid powder that ranges from near colorless to white and may cause eye, nose, and throat irritation.

Eucalyptol (aka 1,8 cineole)

This terpene is named after the tree in which it is found to be most abundant, the eucalyptus tree. It is used in insect repellents and insecticides, and frequently used in cosmetics (e.g., mouthwash) and chemical industries (e.g., cleaning agents) as a flavor and scenting agent. It has a mint-like cooling scent.

Summary of selected effects on the body: Reduces phlegm, cough, and nasal congestion; produces analgesia; may be protective against certain cancers (leukemia).

Eucalyptol is primarily used in medicinal products aimed to mitigate symptoms of **cough, nasal congestion,**[1] **asthma,**[2,3] inflammation-based muscular pain,[4] and rheumatism pain.

A laboratory experiment conducted by a Japanese university in 2002 discovered that the terpene reduced **leukemia** cancer cells by inducing apoptosis (instructing the cancer cell to commit suicide).[5]

Summary of selected effects on mind and emotion: Better concentration, improved cognition, mood improvements.

An experiment conducted on 24 healthy human volunteers demonstrated a detectable concentration of 1,8 cineole (eucalyptol) in the blood of the participants. Blood plasma concentrations of eucalyptol corresponded to improved cognitive performances in speed and accuracy after inhalation of essential oil of rosemary rich in eucalyptol.[6]

Another study looking at the effects of inhaled essential oil of rosemary discovered measurable stimulatory changes in autonomic functions such as blood pressure, heart rate, respiratory rate, and electroencephalography (EEG). Subjects became more active and felt an **uplift in mood** (e.g., felt fresher mentally, less drowsy, more active, less bored).[7]

Safety: Isolated eucalyptol (C10H18O) is a colorless liquid that has been given GRAS (generally recognized as safe) status.

Geraniol

This terpene has a rose-reminiscent scent that is often employed in the creation of perfumes and in cigarettes. It has been shown to be a functional insect repellent against mosquitoes and to attract beneficial insects such as honeybees.

Summary of selected effects on the body: Antioxidant, neuroprotective, possible inducing vasorelaxation.

Researchers discovered that geraniol **reduces oxidative stress,** mitochondrial dysfunction, and neurotoxicity in a disease model using fruit flies that might have relevance to patients with **neuropathies** common in diabetes.[1] Another experiment demonstrated that geraniol induced a relaxant effect on the aorta (the largest blood vessel in the mammalian body) of rats by counteracting the effects of noradrenaline.[2]

Summary of selected effects on mind and emotion: None known.

Safety: Concentrated geraniol (C10H18O) is a colorless to pale yellow oily liquid and is considered an allergen that may cause skin, eye, and respiratory irritations.

Limonene

Limonene is the second-most abundant terpene in nature after pinene. It takes its name from the lemony, orange, citrus-like scent that allows your nose to quickly identify its presence. Upon inhalation, limonene becomes rapidly bioavailable in humans.[1]

Summary of selected effects on the body: May protect against breast cancer, antioxidant, may reduce obesity and reduce weight loss, antifungal, may help with cholesterol-based gallbladder stones, may normalize peristalsis in cases of heartburn and GERD.

D-limonene is highly lipophilic and as a result it spreads readily, especially into adipose (fat) and breast tissues in humans, where it concentrates. It stays bioavailable after oral consumption and may be helpful in reducing breast cancer.[2] Similarly, dietary limonene has demonstrated anti-carcinogenic effects in cases of prostate cancers.[3]

Limonene was also found to be effective against fungi (**antifungal**) responsible for skin infections.[4]

This lemony monoterpene has been reported to dissolve cholesterol-based **gallbladder** stones, reduce stomach acidity, and to normalize peristalsis (**gastrointestinal problems**) relevant to patients suffering from heartburn and GERD.[5]

D-limonene demonstrated significant **anti-inflammatory** properties for in vitro and in vivo studies.[6]

Animal studies revealed that d-limonene should be considered as a possible agent to reduce the deleterious effects of unhealthy bad-fat diets, and to function as a potent **antioxidant** and substance with **blood-pressure** lowering properties.[7] It can also aid in reducing **obesity** and achieving **weight loss**.[8]

Summary of selected effects on mind and emotion: Uplifting, antidepressant, anti-stress, relaxing, anxiolytic, calming effect may help with insomnia.

A study conducted on 12 patients suffering from **depression** discovered that "the treatment with citrus fragrance normalized neuroendocrine hormone levels and immune function and was rather more effective than antidepressants."[9] In addition, numerous pre-clinical and clinical models demonstrated that d-limonene could produce **anti-stress** effects via modulation of parasympathetic and central neurotransmitter functions.[10] Findings from another experiment conducted on mice suggest that a mixture of cis and trans (+)-limonene epoxide **reduced anxiety** without any observable adverse effects.[11] A laboratory study discovered that d-limonene directly binds to the adenosine receptor, which may induce **sedative effects** and thus may be useful in cases of **insomnia**.[12]

Safety: The terpene limonene (C10H16) is a clear, colorless liquid and has been given GRAS (generally recognized as safe) status.

Linalool

Linalool is a simple terpene that offers a floral, lavender-like scent with a hint of spiciness. In fact, the major terpenes that are produced by lavender[1]—namely, linalool (26%), caryophyllene (8%), and myrcene(5%)—are also major terpenes found in cannabis.

Summary of selected effects on the body: Anticonvulsant, antiparasitical (leishmania), may reduce high blood pressure, may produce synergistic analgesic effects with opioids, may protect against certain cancers (kidney, breast, lymphoma, leukemia) and may produce a synergistic effect with certain chemotherapy drugs.

Researchers have demonstrated that linalool produces **anticonvulsant,** anti-glutamatergic,[2] and **anti-leishmanial** (anti-parasite) activities.[3] Linalool has also been found to **reduce blood pressure** in rats.[4] One placebo-controlled study conducted on 54 human patients undergoing the same type of stomach surgery discovered that those patients who were given a "drop" of essential oil of lavender (rich in linalool) needed significantly less opioids to achieve the same **analgesic** effects.[5]

A laboratory study discovered that linalool demonstrated an anticancer effect on renal cell adenocarcinoma cells (**kidney cancer**) comparable to the commercial chemotherapy drug Vinblastine.[6] Another laboratory test suggested that linalool may synergize the therapeutic effect of anthracyclines (chemotherapy drugs) in the management of **breast cancer**, especially in multi-drug-resistant tumors.[7]

Linalool has a strong activity against certain **lymphoma** and **leukemia** cancer cell lines in the petri dish.[8]

An experiment using a mouse model of **Alzheimer's disease** discovered that orally linalool-treated mice (3× daily) reversed the cellular pathological hallmarks of AD, and the treatment restored mental and emotional functions via an **anti-inflammatory** effect.[9] In what is perhaps good news for people who like to inhale, linalool demonstrated a significant protective anti-inflammatory effect in response to the distress produced by cigarette smoke.[10]

Summary of selected effects on mind and emotion: Relaxing, calming, anxiolytic.

Mice that inhaled linalool from ambient air for one hour showed a significant reduction in motility, suggesting a significant **sedative effect.**[11] In addition to its calming effect, linalool reduces anxiety.[12]

Safety: The isolated terpene (C10H18O) is a colorless liquid and may cause skin, eye, and respiratory irritation, and allergic reactions.

Myrcene

Myrcene is a monoterpene that smells like a mix of herbal, woody, and terpy, with undertones of rose and carrot. In very diluted forms the taste of mango is commonly described. Myrcene makes THC more biologically available by increasing the saturation level and speed at CB1 and thus boosting THC's properties. Higher concentrations of beta-myrcene in cannabis tend to produce the feelings of heavy, deep relaxation. Both sativa and indica strains tend to contain myrcene, with indicas often containing more.

Summary of selected effects on the body: Possible sleeping aid, analgesia, may prevent peptic ulcer diseases.

Myrcene induced muscle-relaxant effects and increased barbiturate-induced sleep time in mice,[1] indicating its potential value as a **sedative** and **sleeping aid** in cases of **insomnia.**

Studies conducted on rodents suggested that myrcene induced potent **analgesic effects.**[2,3]

Research also discovered that beta-myrcene may play a role in preventing **peptic ulcer diseases.**[4]

Summary of selected effects on mind and emotion: Sedating, heavy, sleepy, couch-lock effect, relaxing.

Safety: The isolated terpene (C10H16) is a colorless clear liquid that may cause skin, eye, and respiratory irritation.

Pinene

The single terpene that stands out from the crowd with its sheer abundance in the fauna and flora of nature is alpha pinene. It is the terpene that produces the pine-like, woody scent in numerous plants and trees of great variety and from very different geographic regions. Pinene-rich plants and trees include rosemary, pine, cedar, redwood, and other conifers, for example.

Summary of selected effects on the body: Bronchodilation, antifungal (candida), anti-bacterial (MRSA), may help Alzheimer's disease, Parkinson's disease, and multiple sclerosis, may inhibit liver cancer.

Pinene has been found to dilate the bronchi (**bronchodilation**) in human subjects, even at low concentrations.[1] Research tests discovered that only the positive (but not negative) enantiomers of the (+)α- and (+)beta-isomers of pinene were active against fungi and bacteria such as *Candida albicans* and methicillin-resistant *Staphylococcus aureus* or **MRSA (antibiotic-resistant strains).**[2]

Pinene has been identified as an inhibitor of the enzyme that breaks down acetylcholine, the neurotransmitter associated with memory and remembering, and thus increases its bioavailability,[3] making pinene potentially relevant for patient pathologies with **low acetylcholine** such as **Alzheimer's disease, Parkinson's disease**, and **multiple sclerosis.**

Alpha-pinene was found to significantly inhibit liver cancer cell growth in experiments both in vitro (petri dish) and in vivo (on mice).[4]

Summary of selected effects on mind and emotion: Focus, alertness, improved memory.

Pinene may also be of relevance to patients who are concerned with **short-term memory deficits,** a potential adverse effect of THC.

Safety: Isolated pinene (C10H16) is a colorless to pale yellow liquid that may cause irritation to eyes, skin, and mucous membranes.

Terpineol

Similar to terpinolene, terpineol has a scent mixture of pine, lilac, citrus, woody, and floral tones that is commonly employed as a scent and flavoring agent in foods, perfumes, cosmetics, and cleaning products. Alpha-terpineol has shown to enhance the permeability of skin to lipid-soluble compounds,[1] which may help create a topical entourage effect with lipophilic cannabinoids. For instance, if a patient would like to employ the anti-acne effects of the lipophilic cannabinoid CBD, combining it with a small amount of terpineol may enhance localized bioavalability and better possible therapeutic results.

Summary of selected effects on the body: May be able to destroy certain cancers such as lung cancer cells and leukemia cells (erythroleukemic K562).

Swedish researchers discovered that alpha-terpineol was able to destroy lung cancer cell lines in the laboratory.[2] Additionally, Italian oncologists discovered that alpha-terpinol showed important antiproliferative effects of a very specific form of leukemia cells (erythroleukemic K562).[3]

Summary of selected effects on mind and emotion: Calming, relaxing.

Safety: Isolated alpha-terpineol (C10H18O) appearance ranges from a colorless viscous liquid to a solid and may cause irritation to skin, eyes, and mucous membranes.

Terpinolene

Depending on concentration, terpinolene has been described as a fresh, sweet, woody, pine, or floral scent.

Summary of selected effects on the body: Sedative.

One study conducted on mice discovered that inhaled essential oil of *Microtoena patchoulii* leaves produced a **sedative effect,** and that terpinolene, its active ingredient, was essential to produce sedation.[1]

Increased expression and/or activation of protein kinase AKT is involved in a variety of human cancers. A laboratory experiment discovered that terpinolene significantly reduced AKT1 in certain cancer cells and reduced their proliferation.[2]

Summary of selected effects on mind and emotion: Calming, sleepiness, sedative effects.

Safety: Terpinolene (C10H16) is a clear and colorless liquid that has been approved by the FDA as a food additive for human consumption (FDA PART 172).

Sesquiterpenes: Beta-Caryophyllene, Caryophyllene oxide, Humulene, Nerolidol

Beta-Caryophyllene

Beta-caryophyllene is a sesquiterpene that binds with CB2 receptors. Depending on concentration, its scent is described as spicy, peppery, with a strong warm, wooden aroma and an undertone of citrus. It is not uncommon for beta-caryophyllene to be the dominant terpenoid in cannabis.

Summary of selected effects on the body: Immune-modulation, anti-inflammatory, neuroprotective (spasms, seizures, epilepsy, induces neuritogenesis); may protect from Alzheimer's disease and vascular dementia; may be beneficial in cases of colitis; anti-bacterial (tuberculosis, leprosy, *Staphylococcus aureus, Pseudomonas aeruginosa*); anti-fungal (*Candida albicans, Aspergillus niger*);

may be protective against Type II diabetes; may have anti-cancer properties (kidney, colon, pancreatic, melanoma, lymphoma); may produce a synergistic effect with certain chemotherapy agents (e.g., Paclitaxel), analgesic effects (in cases of inflammatory and neuropathic pain), antioxidant, may protect against binge drinking or chronic alcohol consumption–induced alcoholic steatohepatitis, nonalcoholic fatty liver disease, and steatosis (fatty liver diseases).

Beta-caryophyllene binds to CB2 with a Ki of 155 ± 4 nM, while THC binds to the same receptors with a Ki of 35 nM. Thus (E)⬚BCP) has about one fifth of the strength of THC to initiate the same general biological responses such as **anti-inflammatory** and **immune-modulation effects.**[1]

Beta-caryophyllene may hold specific value for patients with certain **neurological** conditions such as **spasm, convulsion,** and **epilepsy.** In pre-clinical trials it displayed anti-convulsant or anti-seizure activity without adverse effects. The anticonvulsant dose of beta-caryophyllene used in the experiment (conducted on mice) was 100 mg/kg.[2] In addition, beta-caryophyllene **induces neuritogenesis** (facilitates the growth of the projections of nerve cell such as dendrites and axons). While the terpenoid is **neuroprotective** and **anti-inflammatory** via CB2 activation, this newly discovered property was induced via tyronsine kinase receptors and thus by a new mechanism independent of CB2.[3]

Moreover, beta-caryophyllene has shown that CB2 and PPARγ receptor activation reduces neuro-inflammatory response in **Alzheimer's disease.**[4] Beta-caryophyllene also reduces cognitive deficits via CB2 activation in a similarly detrimental condition, that of **vascular dementia.**[5]

While the general anti-inflammatory properties of beta-caryophyllene are well known,[6] information about specific **inflammatory conditions** that may benefit from beta-caryophyllene is based in part on pre-clinical trials that determined that beta-caryophyllene may provide a novel treatment possibility in cases of **colitis** via initiating therapeutic actions via CB2 and PPARγ receptor pathways.[7] Another study suggests that beta-caryophyllene modulates the inflammatory reaction caused by the bacteria responsible for **tuberculosis** and **leprosy.**[8] The **antimicrobial** properties of this terpenoid were also effective against *Staphylococcus aureus.*[9] Additional research reveals that beta-caryophyllene showed significant inhibitory effects against bacteria and fungi such as *Pseudomonas aeruginosa,* a bacteria that causes **urinary tract infections,** airway infection, or systemic infections, especially in patients with a weak immune system. ß-caryophyllene is also effective against *Candida albicans* (fungal infections) and *Aspergillus niger* (fungal-based lung infection).[10]

Beta-caryophyllene plays a regulatory role in glucose-stimulated insulin release via CB2 receptor activation, and thus may be protective in **Type II diabetes** via CB2.[11]

The **anticancer** properties of beta-caryophyllene have been demonstrated in the laboratory against **colon cancer** cell lines and **pancreatic cancer** cells,[12] **melanoma** (skin cancer), and lymphoma cells.[13] The anticancer effects were confirmed by another laboratory experiment using several different human cancer lines.[14]

While the potential anticancer properties of THC and CBD have been elucidated in greater detail, the exact mechanism by which these BCP-containing substances produce anticancer properties is yet to be fully understood. Researchers posit that BCP has the potential to suppress the proliferation of cancer cells and thus to induce apoptosis (self-destruction of cancer cells).[15]

Beta-caryophyllene has been found to **potentiate** the effects of other anticancer agents such as the chemotherapy drug **Paclitaxel.** Researchers reported that the terpenoid increased the anticancer effect of Paclitaxel 10-fold.[16] In line with this finding, another laboratory study discovered that beta-caryophyllene demonstrated an anticancer effect on renal cell adenocarcinoma cells (**kidney cancer**) comparable to that of the commercial chemotherapy drug **Vinblastine.**[17]

With regard to **pain,** research on mice has shown that beta-caryophyllene produces **analgesia** via CB2 receptor activation in cases of **inflammatory** and **neuropathic pain** models.[18] Taking the facts into account that in most cases cancer and pain go hand in hand, and that beta-caryophyllene has demonstrated potent analgesic properties as well as being anticancer with a good safety record, it would appear to have potentially therapeutic utility in oncology.[19]

Oxidative stress is a common mechanism making a great variety of chronic conditions worse. beta-caryophyllene has been shown to have strong **antioxidant** effects.[20]

Beta-caryophyllene may be of relevance to numerous liver disorders. For instance, it may prevent and reduce **nonalcoholic fatty liver disease** and its associated metabolic disorders.[21] It may also prevent liver injury from oxidative stress, inflammation, and steatosis (**fatty liver diseases**).[22] Additionally, a study conducted on mice discovered that beta-caryophyllene protects against **binge drinking** or **chronic alcohol consumption**–induced alcoholic steatohepatitis by attenuating inflammation and metabolic dysregulation.[23]

Pet lovers and veterinarians take note: Beta-caryophyllene has **antimicrobial** activity **against dogs' dental plaque-forming bacteria**, thus providing a novel alternative to prevent and treat periodontal disease in dogs.[24]

Summary of selected effects on mind and emotion: Mood improvements (anxiolytic, anti-depressant), no mind-altering effects.

Pre-clinical trials conducted on mice are suggesting that beta-caryophyllene may have CB2-initiated anxiolytic and anti-depressant effects. If these effects are confirmed in humans, this terpene may be relevant in cases of mood disorders.[25]

Safety: Beta-caryophyllene ($C15H24$) is a colorless and slightly oily liquid that has been approved by the FDA as a food additive for human consumption (FDA PART 172).

Caryophyllene Oxide

Caryophyllene oxide is produced when beta-caryophyllene binds with an oxygen atom. Both are sesquiterpenes. Interestingly, researchers demonstrated that beta-caryophyllene-oxide-induced effects, while in many ways mirroring that of beta-caryophyllene, do not occur via binding with CB2 receptor sites and thus must exert their therapeutic powers in a yet-to-be-discovered form.[1]

Summary of selected effects on the body: Possible anticancer properties (cervical, leukemia, lung, gastric, ovarian, breast), may be synergistic with chemotherapy drugs; antifungal (onychomycosis).

The anticancer properties of beta-caryophyllene oxide have been demonstrated in the petri dish against various cancer cell lines such as human **cervical adenocarcinoma** cells, human **leukemia** cancer cells, human **lung cancer** cells, and human **gastric cancer** (stomach) cells.[2] Moreover, another study revealed moderate anticancer properties of beta-caryophyllene oxide against human **ovarian cancer** cells.[3] In addition, a laboratory test demonstrated caryophyllene oxide's synergistic potential by increasing the cellular permeability of the chemotherapy drug paclitaxel and thus potentiating (about 10-fold) its effectiveness in the elimination of breast cancers cells.[4]

Research findings suggest that beta-caryophyllene oxide is able to interfere with multiple mechanisms involved in the production of cancer and thus may be a potential therapeutic candidate for both prevention and treatment. More specifically, beta-caryophyllene oxide stimulated reactive oxygen species production from within the mitochondria, which in turn produced apoptosis. In the same experiment it was discovered that beta-caryophyllene oxide also interferes with several other mechanisms of cancer production such as cell proliferation, survival, angiogenesis, and metastasis of tumor cells. It is interesting to note that similar to beta-caryophyllene, the oxide version **potentiates the anticancer effects of pharmaceutical drugs** aimed at the same targets.[5]

Foot nail fungus is a stubborn and pernicious disease of a chronic nature. Beta-caryophyllene oxide has been found to be extremely effective against **onychomycosis**.[6]

Summary of selected effects on mind and emotion: None.

Safety: Isolated beta-caryophyllene oxide ($C15H24O$) most commonly appears as a pale, yellow-white crystalline solid and is described as having a sweet, fresh, dry, woody, spicy aroma. It has been approved by the FDA as a food additive for human consumption (FDA PART 172).

Humulene

Humulene is a sesquiterpene that smells like hops (used in the brewing of beer) or cannabis.

Summary of selected effects on the body: Anti-inflammatory.

Research conducted on mice discovered that alpha-humulene is readily absorbed (topically and orally) and provides rapid onset of systemic and topical anti-inflammatory properties.[1]

Summary of selected effects on mind and emotion: None known.

Safety: Humulene ($C15H24$) appears as a pale liquid with colors ranging between yellow, green, and clear. In its concentrated form it has a woody scent. In its concentrated form it may cause irritation to skin, eye, and mucous membranes including the respiratory tract.

Nerolidol

Nerolidol has a fresh and woody scent. Poor skin penetration limits the bioavailability of drugs used to treat certain conditions, especially those of the skin. Similar to dimethyl sulfoxide, nerolidol has the capacity to **enhance skin penetration** and thus may be useful in focusing the delivery of drugs where they are needed.

Summary of selected effects on the body: Sedative, antispasmodic, antifungal (*Microsporum gypseum*), antiparasitical (babesiosis, malaria), may produce a synergy with pharmaceutical antibiotics.

As early as 1972, French researchers described the **sedative** and **antispasmodic** properties of nerolidol.[1]

A topical application of nerolidol was tested on guinea pigs and found clinically effective against skin lesions infected with *Microsporum gypseum*[2]; thus it may have **antifungal properties.** A laboratory study showed that certain parasites (e.g., those causing the **tick-borne parasitical illness**

babesiosis) were unable to grow in petri dishes treated with the terpene.[3] In addition, nerolidol was found to be a very effective **anti-malarial** terpene with 100% growth inhibition in the petri dish.[4] Nerolidol has also shown the ability to enhance bacterial cell permeability and thus make bacteria more susceptible to common antibiotics (**antibacterial synergy**), creating a potential basis for overcoming increasing bacterial drug resistance.[5]

Summary of selected effects on mind and emotion: Possibly mildly sedating.

Safety: Nerolidol ($C_{15}H_{26}O$) appears as an oily liquid ranging in color from colorless to yellow. At full strength the scent has been described as floral, lemony, woody. It has been approved by the FDA as a food additive for human consumption (FDA PART 172).

Diterpenes: Phytol

Phytol

Phytol is a terpene that is richly represented in nature as part of the larger molecule chlorophyll, which gives leaves their green color and which is essential in photosynthesis. As chlorophyll breaks down, phytol is released. It is estimated that about 30% of chlorophyll is phytol.

Summary of selected effects on the body: Possible analgesic, antioxidant, anti-inflammatory, antimicrobial (*Staphylococcus aureus*), antiparasitical (schistosomiasis).

Cannabis-based phytol scent is described as a mild aroma reminiscent of older green jasmine tea. Phytol may show promise as a novel agent in the treatment of **pain** and **oxidative stress**.[1] These potential properties were confirmed by researchers from Brazil who discovered that phytol reduces pain perception via reducing inflammatory responses by modulating pro-inflammatory cytokines and oxidative stress. These results suggest a possible role for phytol as an **anti-inflammatory** agent.[2]

Phytol slows the growth of *Staphylococcus aureus* and thus may also provide **antimicrobial** properties.[3]

Phytol may emerge as a new and potentially potent agent in the fight against an infectious parasitical disease called **schistosomiasis**.[4] This parasite infects hundreds of millions of people worldwide and is spread by swimming in freshwater. Infected freshwater snails spread the parasite into the water supply, where it is readily picked up by people swimming, bathing, or washing clothes. To date, only one pharmaceutical drug, Praziquantel, is used in the treatment of the disease. While significantly less expensive in developing countries, for this drug in the US a course of treatment may cost hundreds of dollars.

Also, real concerns about the worm developing drug resistance make it urgent to find inexpensive, safe, and effective alternatives. Phytol may fit the bill. It is a common food additive with a good safety record. Research conducted on mice discovered that a single dose of phytol (40 mg/kg) significantly reduced the presence of the worms and eggs in the blood of the infected animals.[5]

Summary of selected effects on mind and emotion: Relaxing effects.

Since phytol has been shown to be able to inhibit the enzyme that breaks down GABA (a natural neurotransmitter responsible for calming effects), it may increase GABA's bioavailability and produce mildly **relaxing effects**.[6]

Safety: Phytol ($C_{20}H_{40}O$) is a clear, colorless, oily liquid. In concentrated form it has been described as having a subtle balsamic scent.

Major Cannabinoids (in Alphabetical Order)

Cannabichromene (CBC)

CBC is a very stable cannabinoid that has been found in century-old samples of cannabis.[1] When thoroughly purified CBC does not show any activity related to activation of CB1,[2] however, CBC is a potent activator of TRPA1.[3]

Summary of selected effects on the body and synergistic terpenoids: Antidepressant, antibacterial.

CBC displayed significant **antidepressant-like** effect in models of behavioral despair.[4]—eucalyptol (aka 1,8 cineole), limonene, beta-caryophyllene.

CBC demonstrates potent antibacterial activity against MRSA.[5]—pinene, phytol, beta-caryophyllene, pinene.

Summary of selected effects on mind and emotion: None known.

Cannabidiol (CBD)

The actions of CBD all by itself are very complex; however, in combination with other cannabis constituents, it holds even more potentially therapeutic properties. For a more detailed description of CBD and its entourage effects with THC, see chapter 4 or review the CBD summary effects chart in chapter 3.

Summary of selected effects on the body and synergistic terpenoids: Alzheimer's sisease, Parkinson's disease, synergistic analgesia with opioids, analgesia in neuropathies, antibacterial (MRSA), may be antiproliferative effects against cancer (lung cancer, breast cancer), analgesia in neuropathies, acne, antioxidant.

Potentially protective against **Alzheimer's disease** (via GPR3)[1]—pinene.

Potentially protective against **Parkinson's disease** (via GPR6)[2]—pinene.

Potentially synergistic **analgesia** with opioids[3]—linalool; other analgesic terpenes include beta-caryophyllene, borneol, camphene, eucalyptol (aka 1,8 cineole), myrcene, phytol.

Potent **antibacterial** activity against MRSA[4]— pinene, phytol, beta-caryophyllene, pinene.

CBD regulates gene expression (via PPARs) and has been shown to inhibit **lung cancer.**[5]—alpha-terpineol.

Potentially protective against **breast cancer**—limonene, beta-caryophyllene oxide.

Analgesic in cases of acute inflammation via TRPV1.[6]

CBD binds with TRPV1 receptors[7,8] and produces relief in cases of **neuropathies**[9]—geraniol.

CBD may protect against **acne** breakouts.[10]—alpha-terpineol has shown to enhance the permeability of skin to lipid-soluble compounds.[11] CBD is lipid-soluble (lipophilic).—Also, consider limonene, beta-caryophyllene

Studies have shown the effects of CBD to be **neuroprotective** and **antioxidant,** which makes it of value to numerous patient populations, especially those suffering from chronic diseases with underlying pathologies of oxidative stress, neurodegenerative illnesses, aging, ischemia, inflammatory states, or immune disorders for instance.[12-17]—geraniol, beta-caryophyllene. Antioxidant[18]—limonene, geraniol, beta-caryophyllene, phytol.

CBD is a potent anticonvulsant for various disorders involving, spasms, fits, convulsions, seizures, and various forms of epilepsy.[19-24]—beta-caryophyllene.

Summary of selected effects on mind and emotion: Mood improvements, anxiolytic, anti-depressant, antipsychotic, increases libido, anti-anxiety.

CBD activates 5HT1A receptors[25] and thus lowers heart rate[26]—geraniol (vasorelaxation)

mood improvements[27]—beta-caryophyllene, eucalyptol (aka 1,8 cineole)

arousal[28]—consider geraniol (vasorelaxation), limonene for relaxing and uplifting effects.

CBD, especially in the form of tinctures, has become a popular anxiolytic in adults and pediatric patients alike.[29-32]—limonene, linalool, beta-caryophyllene.

CBDA (Cannabidiolic acid)

Anti-emetic (reduces nausea and vomiting)[33-35]—THCA, THC.

Anticancer (reducing breast cancer cell migration)[36]—CBD, beta-caryophyllene, limonene, linalool, eucalyptol, pinene, terpineol.

Cannabigerol (CBG)

Cannabigerol has a relatively low affinity for CB1, CB2; however, CBG is a significant agonist for TRPV1 and TRPA1, and an inhibitor (antagonist) of TRPM8.[1] In addition, CBG is a potent activator of α-2 adrenergic receptors[2] and in contrast to CBD, CBG is a moderate inhibitor of 5HT1A serotonin receptors.[3]

Summary of selected effects on the body and synergistic terpenoids: Muscle relaxant, analgesia, sedative, may be antiproliferative in cases of prostate cancer, antibacterial (MRSA), appetite stimulant (in cases of anorexia, cachexia), anti-parasitical (leishmania).

CBG-based activation of α-2 receptors inhibits catecholamine release thus enhances muscle relaxation, analgesia, and sedation.[4]

Muscle relaxation—geraniol (vasorelaxation).

Analgesia—linalool, beta-caryophyllene, borneol, camphene, eucalyptol (aka 1,8 cineole), myrcene, phytol.

Sedation—myrcene, terpenolene, nerolidol.

Potent **antibacterial** activity against MRSA[5]—pinene, phytol, beta-caryophyllene, pinene.

CBG is a potent antagonist of the menthol receptor TRPM8, a target of relevance for **prostate cancer**[6]—limonene.

CBG is an appetite stimulant and thus may help with **anorexia** and **cachexia**[7]—there is no known terpene synergy to enhance one's appetite (humulene is an appetite suppressant).

CBG has produced antileishmanial activities.[8]—linalool.

Summary of selected effects on mind and emotion: Unknown.

Cannabinol (CBN)

Most CBN found in cannabis plant material is due to chemical changes that naturally occur during the drying process. As THC begins to degrade over time, some of its chemical composition transforms into CBN. Research seems to indicate that CBN promotes sedation even at low concentrations. CBN is not a mind-altering cannabinoid or at best is a very mild psychoactive compound.

Selected CBN induced effects on the body and synergistic terpenoids: Antibacterial (MRSA).

Potent **antibacterial** activity against MRSA[1]—pinene, phytol, beta-caryophyllene.

Summary of selected effects on mind and emotion: Relaxation, sedation, and sleep.

CBN may promote relaxation, sedation, and sleep.[2,3]—myrcene.

Δ9-Tetrahydrocannabinol (THC)

THC is at the center of understanding cannabinoid synergies. And, while THC is the single most investigated cannabinoid, it is continuing to surprise researchers as new even more complex information is revealed regarding the many ways it affects human biology as well as psychology. For more in-depth information, see chapter 3, section THC.

Summary of selected effects on the body and synergistic terpenoids: Analgesia, immunomodulation, bronchiodilation, relaxant (anti-stress), anti-emetic, appetite stimulant.

Analgesia[1,2]—beta-caryophyllene.

Immunomodulation[3-5]—beta-caryophyllene.

Bronchiodilation[6-9]—pinene.

Appetite stimulation[10,11]—CBG.

Anti-emetic[12]—THCA.

Summary of selected effects on mind and emotion: Stress reduction, relaxing.

One of the key attributes of THC is its capacity to mitigate the impact of **excess stress**[13]—myrcene.

Tetrahydrocannabinolic acid (THCA)

THCA is not psychoactive in fresh, raw cannabis, so this may be an ideal option for patients who need or want to avoid mind-altering effects while engaging the raw plant's potentially therapeutic properties. Research in this area is at an early stage, but results indicate a possible role for THCA in cases of

Anti-emetic (reduces nausea and vomiting)[14]—CBDA, THC.

Neuroprotective[15]—beta-caryophyllene, pinene, geraniol.

Anti-inflammatory[16]—beta-caryophyllene, phytol, myrcene, humulene.

Top 10 Conditions:
Recent Informative or Clinical Trials for Deeper Learning

1. Cancer (Brain Cancer)

Schloss J, Lacey J, Sinclair J, Steel A, Sughrue M, Sibbritt D, Teo C. A Phase 2 Randomised Clinical Trial Assessing the Tolerability of Two Different Ratios of Medicinal Cannabis in Patients With High Grade Gliomas. Front Oncol. 2021 May 21;11:649555.

2. Heart Disease

Garza-Cervantes JA, Ramos-González M, Lozano O, Jerjes-Sánchez C, García-Rivas G. Therapeutic Applications of Cannabinoids in Cardiomyopathy and Heart Failure. Oxid Med Cell Longev. 2020 Oct 27;2020:4587024.

3. Chronic Pain (Non-Malignant)

Jeddi HM, Busse JW, Sadeghirad B, Levine M, Zoratti MJ, Wang L, Noori A, Couban RJ, Tarride JE. Cannabis for medical use versus opioids for chronic non-cancer pain: a systematic review and network meta-analysis of randomised clinical trials. BMJ Open. 2024 Jan 3;14(1):e068182.

4. Diabetes

Jadoon KA, Ratcliffe SH, Barrett DA, Thomas EL, Stott C, Bell JD, O'Sullivan SE, Tan GD. Efficacy and Safety of Cannabidiol and Tetrahydrocannabivarin on Glycemic and Lipid Parameters in Patients With Type 2 Diabetes: A Randomized, Double-Blind, Placebo-Controlled, Parallel Group Pilot Study. Diabetes Care. 2016 Oct;39(10):1777-86.

5. Anxieties

Raymundi AM, da Silva TR, Sohn JMB, Bertoglio LJ, Stern CA. Effects of Δ9-tetrahydrocannabinol on aversive memories and anxiety: a review from human studies. BMC Psychiatry. 2020 Aug 26;20(1):420.

6. Neurological Conditions (Multiple sclerosis)

Nicholas J, Lublin F, Klineova S, Berwaerts J, Chinnapongse R, Checketts D, Javaid S, Steinerman JR. Efficacy of nabiximols oromucosal spray on spasticity in people with multiple sclerosis: Treatment effects on Spasticity Numeric Rating Scale, muscle spasm count, and spastic muscle tone in two randomized clinical trials. Mult Scler Relat Disord. 2023 Jul;75:104745.

7. Insomnia

Saleska JL, Bryant C, Kolobaric A, D'Adamo CR, Colwell CS, Loewy D, Chen J, Pauli EK. The Safety and Comparative Effectiveness of Non-Psychoactive Cannabinoid Formulations for the Improvement of Sleep: A Double-Blinded, Randomized Controlled Trial. J Am Nutr Assoc. 2024 Jan;43(1):1-11.

Ried K, Tamanna T, Matthews S, Sali A. Medicinal cannabis improves sleep in adults with insomnia: a randomized double-blind placebo-controlled crossover study. J Sleep Res. 2023 Jun;32(3):e13793.

8. Irritable Bowel Syndrome (Crohn's Disease)

Naftali T, Bar-Lev Schleider L, Almog S, Meiri D, Konikoff FM. Oral CBD-rich Cannabis Induces Clinical but Not Endoscopic Response in Patients with Crohn's Disease, a Randomised Controlled Trial. J Crohns Colitis. 2021 Nov 8;15(11):1799-1806.

9. Cough

Menezes PMN, Pereira ECV, Lima KSB, Silva BAOD, Brito MC, Araújo TCL, Neto JA, Ribeiro LAA, Silva FS, Rolim LA. Chemical Analysis by LC-MS of Cannabis sativa Root Samples from Northeast Brazil and Evaluation of Antitussive and Expectorant Activities. Planta Med. 2022 Oct;88(13):1223-1232.

10. Skin Conditions (Acne)

Yoo, Eun Hee, and Ji Hyun Lee. 2023. "Cannabinoids and Their Receptors in Skin Diseases" International Journal of Molecular Sciences 24, no. 22: 16523.

Top 14 Terpenes:
Recent Informative or Clinical Trials for Deeper Learning

1. Caryophyllene

Alizadeh S, Djafarian K, Mofidi Nejad M, Yekaninejad MS, Javanbakht MH. The effect of β-caryophyllene on food addiction and its related behaviors: A randomized, double-blind, placebo-controlled trial. Appetite. 2022 Nov 1;178:106160.

2. Limonene

Chebet JJ, Ehiri JE, McClelland DJ, Taren D, Hakim IA. Effect of d-limonene and its derivatives on breast cancer in human trials: a scoping review and narrative synthesis. BMC Cancer. 2021 Aug 6;21(1):902.

3. Linalool

Dos Santos ÉRQ, Maia JGS, Fontes-Júnior EA, do Socorro Ferraz Maia C. Linalool as a Therapeutic and Medicinal Tool in Depression Treatment: A Review. Curr Neuropharmacol. 2022;20(6):1073-1092.

4. Eucalyptol

Mączka W, Duda-Madej A, Górny A, Grabarczyk M, Wińska K. Can Eucalyptol Replace Antibiotics? Molecules. 2021 Aug 14;26(16):4933.

5. Phytol

Alencar MVOB, Islam MT, Ali ES, Santos JVO, Paz MFCJ, Sousa JMC, Dantas SMMM, Mishra SK, Cavalcante AACM. Association of Phytol with Toxic and Cytotoxic Activities in an Antitumoral Perspective: A Meta-Analysis and Systemic Review. Anticancer Agents Med Chem. 2018;18(13):1828-1837.

6. Nerolidol

Chan WK, Tan LT, Chan KG, Lee LH, Goh BH. Nerolidol: A Sesquiterpene Alcohol with Multi-Faceted Pharmacological and Biological Activities. Molecules. 2016 Apr 28;21(5):529.

7. Myrcene

Surendran S, Qassadi F, Surendran G, Lilley D, Heinrich M. Myrcene-What Are the Potential Health Benefits of This Flavouring and Aroma Agent? Front Nutr. 2021 Jul 19;8:699666.

8. Pinene

Nyamwihura RJ, Ogungbe IV. The pinene scaffold: its occurrence, chemistry, synthetic utility, and pharmacological importance. RSC Adv. 2022 Apr 12;12(18):11346-11375.

9. Camphene

Gadotti VM, Huang S, Zamponi GW. The terpenes camphene and alpha-bisabolol inhibit inflammatory and neuropathic pain via Cav3.2 T-type calcium channels. Mol Brain. 2021 Nov 14;14(1):166.

9. Terpineol

Khaleel, Christina, Tabanca, Nurhayat and Buchbauer, Gerhard. "α-Terpineol, a natural monoterpene: A review of its biological properties" Open Chemistry, vol. 16, no. 1, 2018, pp. 349-361.

11. Terpinolene

Menezes IO, Scherf JR, Martins AOBPB, Ramos AGB, Quintans JSS, Coutinho HDM, Ribeiro-Filho J, de Menezes IRA. Biological properties of terpinolene evidenced by in silico, in vitro and in vivo studies: A systematic review. Phytomedicine. 2021 Dec;93:153768.

12. Borneol

Chebet JJ, Ehiri JE, McClelland DJ, Taren D, Hakim IA. Effect of d-limonene and its derivatives on breast cancer in human trials: a scoping review and narrative synthesis. BMC Cancer. 2021 Aug 6;21(1):902.

13. Geraniol

Rizzello F, Ricci C, Scandella M, Cavazza E, Giovanardi E, Valerii MC, Campieri M, Comparone A, De Fazio L, Candela M, Turroni S, Spisni E. Dietary geraniol ameliorates intestinal dysbiosis and relieves symptoms in irritable bowel syndrome patients: a pilot study. BMC Complement Altern Med. 2018 Dec 19;18(1):338.

14. Humulene

Dalavaye N, Nicholas M, Pillai M, Erridge S, Sodergren MH. The Clinical Translation of α-humulene - A Scoping Review. Planta Med. 2024 May 8.

Endnotes; To access all footnotes use this QR code:

Navigate the Latest Research Effortlessly with CannaKeys (CK360) – Your Go-To Resource for Clinicians and Cannabis Coaches

- CK360 analyzes thousands of scientific studies every year
- CK360 is constantly updated
- CK360 saves you time by providing you with what you need to optimize patient outcomes

Uwe Blesching co-created CannaKeys an online platform that helps clinicians and coaches keep up. Updated daily by our team of expert analysts, providing you with the latest research trends and scientific insights.

Support Clinical Decisions with the Latest Evidence

CannaKeys Testimonials

"Cannakeys is a disruptive tool, gathering the most reliable sources of study and research regarding medical cannabis in a very practical, dynamic, and up-to-date way. My feeling is that Cannakeys will become "the PubMed of Endocannabinoid Medicine", everyone's reference source for searching qualified scientific evidence in the field."

–Patricia Montagner, Neurosurgeon, Founder–WeCann Academy

"CannaKeys is the best and most comprehensive site for tracking down the academic literature on cannabis. The menu is easy to navigate and the database very thorough. This platform should be compulsory for all medical practitioners and others wanting to prescribe cannabis to keep them up to date with this increasingly complex literature."

–Mike Barnes, Honorary Professor of Neurological Rehabilitation, Founding Chair–Medical Cannabis Clinicians Society, UK.